Assessments for the Science Process Skills of Inquiry

Primary Grades

Dr. Karen Ostlund
UTeach–College of Natural Sciences
The University of Texas at Austin

Sheryl Mercier
Kings Canyon Unified School District
Dunlap, California

ISBN 0-9658768-2-9

S&K Associates
Squaw Valley, California
Email: skmer5@aol.com

Assessments for the Process Skills of Inquiry
Primary Grades

This book was set in Tekton font by S&K Associates.
The authors and editors are Karen Ostlund and Sheryl Mercier.
Illustrations by Sheryl Mercier.
Printed in the United States of America.

First Printing 2006

TABLE OF CONTENTS

Level 1

Level 2

Level 3

Scientific Investigations

INTRODUCTION

Assessing Science Process Skills of Inquiry

Assessing primary students' science process skills, their ability to conduct science inquiry, is essential for several reasons. First, if students are to increase their ability to engage in inquiry, teachers need more accurate information regarding what students have learned about science process skills. Second, accurate assessment of science process skills at all grade levels provides teachers, administrators, and policy makers with better data for making instructional decisions. While the National Science Education Standards (NRC, 1998) call for teaching science content in greater depth at each grade level, they also recognize that students need to learn the process skills of inquiry associated with doing science at all grade levels. More accurate assessments of science process skills provides teachers, parents and district administrators with data on the effectiveness of the curricula, and policy makers with information about the effectiveness of the inquiry-based instructional approaches recommended for science teaching and learning.

In spite of the fact that curriculum materials and assessment tools exist for science content or concepts at all grade levels, there is very little information on how to assess the science process skills of inquiry that students are expected to learn as a result of science instruction. It is important to note that the science process skill assessments we developed are not designed to evaluate everything a student should know about science. Knowledge of specific science concepts, for example, is not something these performance tasks can assess. Instruments that document learning of specific science content knowledge are generally prepared to reflect a course of study adopted by a school district or state. Many states have or are developing competency-based assessments in science that evaluate science content knowledge.

Frequently, the feedback provided on these types of instruments is "high-stakes" in that it is used as the primary source of data for evaluating student learning and, in some cases, promotion to the next grade. We intentionally wrote our performance tasks to assess science process skills of inquiry in order to capture a different aspect of learning - the growth of an individual

student's knowledge about the processes that we believe is essential to learning about science inquiry. The performance tasks for assessing science process skills we developed are capable of assessing changes in a student's abilities over time rather than assessing learning at just one moment in time.

The hands-on student performance tasks in this book were developed from our efforts to document changes in student learning. Teachers of primary students who use these performance assessments will be able to observe and record information regarding student ability to use science process skills. Repeated use of these performance tasks by a teacher will capture change in a student's science process skill acquisition over time. Our intent was to create assessment tools that would document change in the ability to use the science process skills of inquiry that may be incremental and not easily captured by one administration of a 'paper and pencil' instrument. Finally, the performance task assessments we developed require few changes to the existing science curriculum for a primary teacher or school. That is, we do not want teachers to think that their curriculum must change to address the items on our assessments. Rather, we want teachers using the assessments to decide if the items on our performance tasks caused any changes in student learning.

Process Skills and Intellectual Development

Many children progress through stages of development that reflect increasingly complex ways of representing what they are learning. The teacher's task when using our performance tasks is to assess a student's stage of intellectual development and then expand upon that student's knowledge and abilities so that each student develops competency in specific intellectual abilities required to engage in the science process skills of inquiry. Keep in mind that children learn at different rates and in different ways. Teachers can plan their instruction, both individual and whole class, in response to feedback they receive from analyzing the hands-on performance tasks using the assessment rubrics. These performance assessments also offer teachers feedback on their instruction, feedback that could be used when planning future instruction. The assessments can help teachers develop a system of documenting student progress that includes portfolios of student work to document growth over time, use assessment instruments that can contribute to their ongoing instruction, and serve as a reporting mechanism for parents.

These assessments include the process of science we believe are applicable when learning a wide variety of science content. The processes we identified as necessary for primary students were observing, communicating, estimating, measuring, collecting data, classifying, inferring, predicting, making models, interpreting data, making graphs, hypothesizing, controlling variables, defining operationally, and investigating. From this list we developed a description of what each process skills involved. We did not specify any particular science content since the teachers using these assessments would need to comply with their district or state frameworks. Then we developed a performance task for each science process skill of inquiry with activities students might engage in that would indicate that they have mastered a particular process skill at each developmental level.

Pyramid of Process Skills for Science

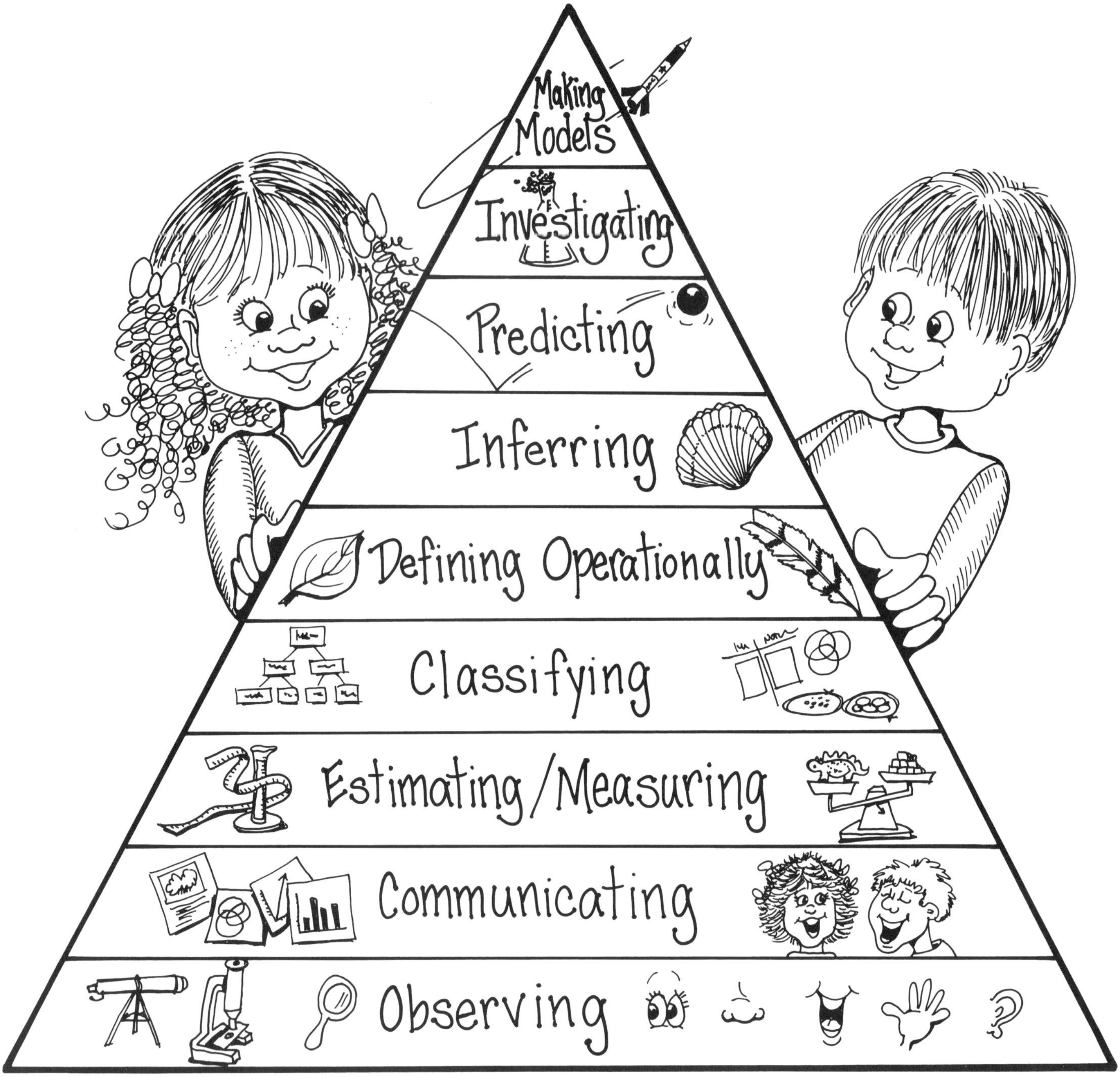

HIERARCHY OF SCIENCE PROCESS SKILLS

The first seven process skills form the foundation for scientific methods. These basic process skills are integrated together when students design and carry out investigations. The seven basic process skills can be put in a logical order of increasing complexity, although even the youngest students will use all of the skills together at various times. In the earliest grades, students spend a large amount of time using skills such as observing and communicating. As students progress to upper grades,r, they will spend more time using the skills of inferring and predicting. Classifying and measuring tend to be used across the grade levels equally.

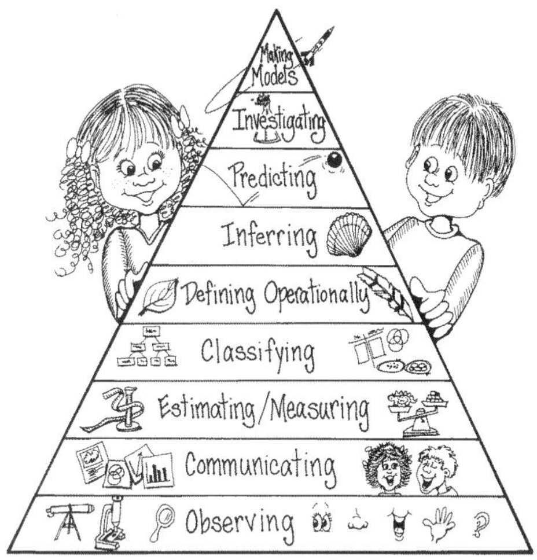

Observing is the fundamental process skill of inquiry. Students observe organisms, objects, and events using their five senses; and this is how they learn about the world around them. The ability to make accurate observations is essential to the development of the other science process skills. Whether students are observing with their five senses or with instruments to aid them, teachers can guide their students to make better and more detailed observations. Teachers can prompt students to add details to their descriptions no matter which of the five senses they are using. *Students should be cautioned that the sense of taste should only be used when the teacher says that it is safe to do so.*

Communicating, the second of the basic science process skills of inquiry, goes hand-in-hand with observing because it involves describing observations using words or graphics to depict an organism, object, or event. Students have to communicate in order to share their observations with someone else, and the communication must be clear and effective. One of the keys to communicating effectively is to use *referents*, i.e., references to items that the other person is already familiar with. The idea is to communicate using descriptive words for which both people share a common understanding. Students can communicate verbally, in writing, drawing pictures or creating graphs, charts, maps, diagrams, or other visuals.

Estimating and Measuring are really just special cases of observing and communicating. Estimating and measuring involve approximating and measuring using both standard and nonstandard units to describe the dimensions of an object or event using observations and communication skills. When students measure some property, they compare the property to a defined referent called a unit. A measurement statement contains two parts, a number to tell us *how much* or *how many*, and a name for the unit to tell us how much *of what*. The use of the number makes a measurement a quantitative observation.

Classifying is the grouping or sequencing of objects or events according to a recognized scheme, based on observations and /or estimating and measuring. Grouping objects is a way of imposing order based on similarities, differences, and interrelationships. The simplest method of classifying is serial ordering in which objects are placed into rank order based on some property. Two other methods of classifying are binary classification and multistage classification. In a binary classification system, a set of objects is simply divided into two subsets. This is usually done on the basis of whether each object has or does not have a particular property. A multi-stage classification is constructed by performing consecutive binary classifications on a set of objects and then on each of the resulting subsets of objects. The result is a classification system consisting of layers or stages. A multi-stage classification is complete when each of the objects in the original set has been separated into a category by itself.

Defining Operationally Defining operationally is the process by which a scientific term is framed in terms of a student's own experiences. An operational definition is also used to define what you will do and what you will observe when conducting an investigation. Defining operationally builds on observing and classifying to develop conceptual understanding.

Inferring requires the use of observations to identify trends, patterns, and cause-and-effect relationships. Unlike observations, which are direct evidence gathered about an organism, object or event, inferences are explanations or interpretations that follow from the observations. Students need to be taught the difference between observations and inferences. They need to be able to differentiate for themselves the evidence they gather about the world as observations and the interpretations or inferences they make based on the observations. Teachers can help students make this distinction by first prompting them to be detailed and descriptive in their observations. Then, by asking students questions about their observations, teachers can encourage the students to think about the meaning of the observations. Thinking about making inferences in this way should remind us that inferences link what has been observed together with what is already known form previous experiences. We use our past experiences to help us interpret our observations.

Predicting is stating the outcome of a future event based on a pattern of evidence obtained through observations, measurements, classification, operational definitions, and inferences. When students make educated guesses about the outcomes of future events, they are forcasting future observations. The ability to make predictions about future events allows us to successfully interact with the environment around us. Predicting is based on good observations and inferences made about observed events. Like inferences, predictions are based on both what we observe and also our past experiences.

Investigating encompasses all the previous process skills in order to ask and answer questions. Investigating includes **hypothesizing** (formulating a testable statement), **controlling variables** (the process by which conditions affecting an event are deliberately controlled and manipulated), **making graphs** (organizing data), and **interpreting data** (drawing conclusions). Science methods are the processes used by scientists who are asking and answering questions by integrating the seven basic science process skills.

Making Models is creating a mental or physical model of a process or event on the basis of acceptable hypotheses that have yet to be tested. Existing models have to be modified to include new observations which sends us down to the bottom of the pyramid to start over again. Mental and physical models are built from successfully integrating the science process skills of inquiry to make learning experiences richer and more meaningful for students. Students will be learning the process skills of science as well as science content. Students will also be actively engaged with the science they are learning and will reach a deeper understanding of the content.

Level 1 Student Name: ______________________ Date: __________

OBSERVING

Materials

Observing Assessment Sheet
Crayons
4 pairs of buttons (total of 12) in a re-sealable bag
(Example Set of Buttons: 2 small 2-hole yellow buttons, 2 small 4-hole yellow buttons, 2 large 2-hole yellow buttons, 2 large 4-hole yellow buttons)

Procedure

Instruct students to observe the buttons. Ask students to describe their observations and create a list of words on the board that students can use when they describe their favorite pair of buttons. Explain that for each button there is a matching button. Tell students to place each pair of matching buttons in one of the boxes on the assessment sheet. Then ask the students to trace around the matching buttons, color them, and draw dots to show the holes in each button. Invite students to identify their favorite pair of buttons by writing the box number and then to describe the pair of buttons using the list of words on the board.

Observing Rubric

4 points	exceptionally accurate/reasonable, complete, detailed
3 points	adequately accurate/reasonable, complete, detailed
2 points	somewhat accurate/reasonable, complete, detailed
1 point	not accurate/reasonable, complete, or detailed
0 points	not attempted

Scoring Criteria- Observing	Points 0-4
Followed directions using appropriate safety precautions to complete the performance assessment.	
Observed and matched the buttons.	
Traced, colored, and drew dots to match the actual buttons and holes in the first box.	
Traced, colored, and drew dots to match the actual buttons and holes in the second box.	
Traced, colored, and drew dots to match the actual buttons and holes in the third box.	
Traced, colored, and drew dots to match the actual buttons and holes in the fourth box.	
Identified the box number for a favorite pair of buttons.	
Described a favorite pair of buttons.	
Total	/64

Comments:

Observing

1. Look and match the buttons.
2. Trace one pair of buttons in each box.
3. Color the buttons and draw dots to show the holes.

A.

B.

C.

D.

4. Which pair of buttons is your favorite? Tell about them.

Level 1 Student Name: ________________________________ Date ______________

COMMUNICATING

Materials
Communicating Assessment Sheet
Crayons

Procedure
Ask students to observe one hand and describe the details they see. Write the words and phrases students use to describe their hands on the board. Instruct students to place their hand so it is centered in the middle of the box on the student page. Have students trace around their hand. Then ask students to look at their hand and use colors to draw the details they observe in their hand. Then invite students to tell everything they can about their hand using the words on the board.

Communicating Rubric

4 points	exceptionally accurate/reasonable, complete, detailed
3 points	adequately accurate/reasonable, complete, detailed
2 points	somewhat accurate/reasonable, complete, detailed
1 point	not accurate/reasonable, complete, or detailed
0 point	not attempted

Scoring Criteria-Communicating		Points 0-4
Followed directions using appropriate safety precautions to complete the performance assessment.		
Centered hand in box and carefully traced around it.		
Used color to draw details in the traced hand.		
The traced hand and details matched the actual hand.		
Described the details in the hand.		
	Total	/20

Comments:

Communicating

1. Trace around your hand.

2. Draw in all the details with color.

3. Tell everything you can about your hand.

Level 1 Student Name: ______________________ Date: __________

ESTIMATING

Materials

Estimating Assessment Sheet
centimeter ruler

Procedure

Instruct students to look at a centimeter ruler. Explain that the width of their index finger is about one centimeter wide. Have students compare the width of their index finger to the centimeter ruler. Then ask the students to estimate the length of each pattern in centimeters using one of their index fingers. Tell students to record their estimate in the box next to each line. Then have students orally explain how they estimated.

Estimating Rubric

4 points	exceptionally accurate/reasonable, complete, detailed
3 points	accurate/reasonable, complete, detailed
2 points	slightly accurate/reasonable, complete, detailed
1 point	not accurate/reasonable, complete, or detailed
0 points	not attempted

Scoring Criteria - Estimating	Points 0-4
Followed directions using appropriate safety precautions to complete the performance assessment.	
Used an index finger to estimate how long the first pattern is in centimeters and wrote the estimate in the box.	
Used an index finger to estimate how long the second pattern is in centimeters and wrote the estimate in the box.	
Used an index finger to estimate how long the third pattern is in centimeters and wrote the estimate in the box.	
Used an index finger to estimate how long the fourth pattern is in centimeters and wrote the estimate in the box.	
Used an index finger to estimate how long the fifth pattern is in centimeters and wrote the estimate in the box.	
Used an index finger to estimate how long the sixth pattern is in centimeters and wrote the estimate in the box.	
Explained the procedure used to estimate the number of centimeters for each pattern.	
Total	/ 32

Comments:

Estimating

1. This is the length of one centimeter.
2. Use your fingers to estimate how long each pattern is in centimeters.
3. Write the number in the box.

4. Explain how you estimated.

Level 1 Student Name: ______________________________ Date: __________

MEASURING

Materials

Measuring Assessment Sheet
glue
14 paper strips cut from an assortment of colored paper
(suggested set of strips):
3 strips of different color 4 cm X 1 cm
2 strips of different color 6 cm X 1 cm
4 strips of different color 8cm X 1 cm
5 strips of different color 2 cm X 1 cm

Procedure

Tell the student to place strips of equal length in the same box on the assessment sheet. Then have the student glue down the strips. Ask students to describe how the lengths compare. Check to see if students are able to use comparison words such as longer, shorter, and same length correctly.

Measuring Rubric

4 points	exceptionally accurate/reasonable, complete, detailed
3 points	adequately accurate/reasonable, complete, detailed
2 points	somewhat accurate/reasonable, complete, detailed
1 point	not accurate/reasonable, complete, or detailed
0 points	not attempted

Scoring Criteria-Measuring	Points 0-4
Followed directions using appropriate safety precautions to complete the performance assessment.	
Strips of the same length are matched and glued in Box 1.	
Strips of the same length are matched and glued in Box 2.	
Strips of the same length are matched and glued in Box 3.	
Strips of the same length are matched and glued in Box 4.	
Described how the lengths of the strips compare.	
Total	/ 24

Comments:

Measuring

1. Compare the strips of paper.
2. Put the strips that are the same length in the same box.
3. Glue down the strips.

Box 1

Box 2

Box 3

Box 4

4. Tell how the lengths compare.

Level 1 Student Name: ______________________ Date: __________

COLLECTING DATA

Materials

Level 1 Collecting Data Assessment Sheet
Glue stick
20 colored squares (2 cm X 2 cm) in a re-sealable plastic bag
Example set of squares: 4 red squares, 5 blue squares, 2 yellow squares, 3 green squares, 6 orange squares

Procedure

Have students sort the squares into groups according to color. Then ask students to arrange all the squares of the same color in a continuous column beginning at the bottom of the assessment sheet. Have the students glue the squares down. Then ask students to count the squares in each column and write the number in the circle below each column.

Collecting Data Rubric

4 points	exceptionally accurate/reasonable, complete, detailed
3 points	adequately accurate/reasonable, complete, detailed
2 points	somewhat accurate/reasonable, complete, detailed
1 point	not accurate/reasonable, complete, or detailed
0 points	not attempted

Scoring Criteria- Collecting Data	Points 0-4
Followed directions using appropriate safety precautions to complete the performance assessment.	
Sorted and counted the squares into groups according to color.	
Arranged the squares of the same color in the **first** column beginning at the bottom and then glued the squares in place.	
Arranged the squares of the same color in the **second** column beginning at the bottom and then glued the squares in place.	
Arranged the squares of the same color in the **third** column beginning at the bottom and then glued the squares in place.	
Arranged the squares of the same color in the **fourth** column beginning at the bottom and then glued the squares in place.	
Arranged the squares of the same color in the **fifth** column beginning at the bottom and then glued the squares in place.	
Arranged the squares of the same color in the **sixth** column beginning at the bottom and then glued the squares in place.	
Arranged the squares of the same color in the **seventh** column beginning at the bottom and then glued the squares in place.	
Total	/36

Comments:

Collecting Data

1. Sort the squares by color.
2. Glue the squares that are the same color in the same column.
3. Count the squares. Write the number in the circle.

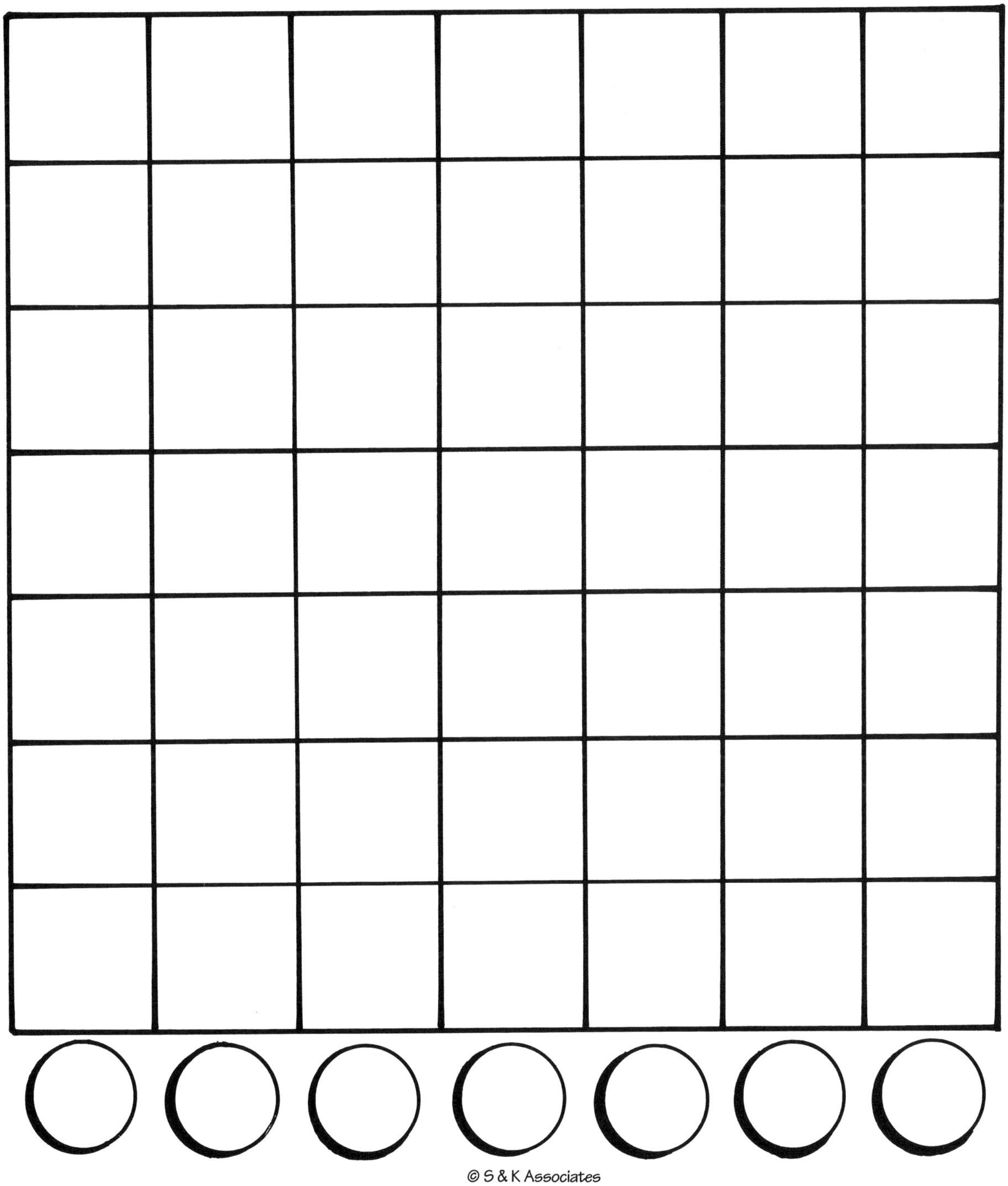

CLASSIFYING

Materials

Classifying Assessment Sheet
crayons
10 small rocks in a re-sealable plastic bag

Procedure

Ask students to observe and describe their rocks. Write the words used to describe the rocks on the board. Then ask students to sort the rocks into two groups by placing the rocks in one of the two the boxes. Tell the students to trace around the rocks in each box and color each tracing to match each rock. Ask students to use the words on the board to describe what is different about the two groups. Then ask students to use the words on the board to describe what is the same about the two groups.

Classifying Rubric

4 points	exceptionally accurate/reasonable, complete, detailed
3 points	adequately accurate/reasonable, complete, detailed
2 points	somewhat accurate/reasonable, complete, detailed
1 point	not accurate/reasonable, complete, or detailed
0 points	not attempted

Scoring Criteria - Classifying	Points 0-4
Followed directions using appropriate safety precautions to complete the performance assessment.	
Observed and sorted the rocks into two groups.	
Traced rocks in **Group 1** and colored the tracings to match each rock.	
Traced rocks in **Group 2** and colored the tracings to match each rock.	
Described what is different about the two groups.	
Described what is the same about the two groups.	
Total	/24

Comments:

Classifying

1. Observe the rocks.
2. Sort the rocks into two groups.

3. Trace around the rocks and color them in.

4. How are the two groups alike? ____________________

____

5. How are the two groups different? ______________

____

Level 1 Student Name: ______________________________ Date: __________

INFERRING

Materials

Inferring Assessment Sheet
Crayons
Sock containing a small object such as a peanut in the shell
Do not use sharp objects,

Procedure

Have each student place one hand into the sock and feel the object. Ask the student to describe how the object feels. Record the responses on the data sheet. (You may want to record words on the board and have students with adequate writing skills write their own responses by using words on the board.) Next, have the student draw a picture of what the object in the sock might look like. After each student draws a picture, allow the students to take the object out of the sock and tell how the object is different from what s/he inferred.

Inferring Rubric

4 points	exceptionally accurate/reasonable, complete, detailed
3 points	adequately accurate/reasonable, complete, detailed
2 points	somewhat accurate/reasonable, complete, detailed
1 point	not accurate/reasonable, complete, or detailed
0 points	not attempted

Scoring Criteria- Inferring	Points 0-4
Followed directions using appropriate safety precautions to complete the performance assessment.	
Described how the object in the sock feels.	
Drew a picture to infer what the object in the sock looks like.	
Described how the actual object is different from the inferred drawing based on how the object felt.	
Total	/16

Comments:

Inferring

1. Put your hand in the sock.
2. Tell how the object feels.

____

____

3. Draw a picture of how you think the object in the sock looks.

4. Look at the object. How is it different from what you

Level 1 Student Name: ______________________________ Date: ______________

PREDICTING

Materials

Predicting Assessment Sheet
5 clear glass soda bottles, water, food coloring, unsharpened pencil

Procedure

Mix water with food coloring so that the color can be seen easily. Then pour the colored water into each of the 5 clear glass bottles so that bottles are 4/5 full, 3/5 full, 2/5 full, and 1/5 full of colored water. Arrange the bottles randomly. Ask the student to identify the bottle with the most water and tap it with the unsharpened pencil to hear pitch (how high or low it sounds). Then ask the student to identify the bottle with the least amount of water and tap this bottle to compare the pitch to the bottle with the most water. Have the student predict how the bottles will sound when tapped by putting the bottles in order from highest to lowest pitch. Then have the student color in the level of the water in the bottles in this order to show the prediction. Instruct the student to test the prediction by tapping each bottle in order to see if they are arranged from highest to lowest pitch. Tell the student to rearrange the bottles until they are in the actual order from highest to lowest pitch. Then have the student color in the bottles to show the level of the water when the bottles are actually arranged from highest to lowest pitch. Have the student compare his/her predicted order to the actual order. Discuss the accuracy of predictions. Explain that one way scientists make a new discovery is when their prediction doesn't match what they think is going to happen.

Predicting Rubric

4 points	exceptionally accurate/reasonable, complete, detailed
3 points	adequately accurate/reasonable, complete, detailed
2 points	somewhat accurate/reasonable, complete, detailed
1 point	not accurate/reasonable, complete, or detailed
0 points	not attempted

Scoring Criteria-Predicting	Points 0-4
Followed directions using appropriate safety precautions to complete the performance assessment.	
Identified the bottle with the most amount of water and tapped it to determine the pitch (how high or low it sounds).	
Identified the bottle with the least amount of water and tapped it to compare the pitch (how high or low it sounds).	
Predicted the order of the bottles from highest to lowest sound by arranging the bottles in order and recorded the prediction by coloring in the level of the water.	
Tapped each bottle to determine if the predicted order of the bottles from highest to lowest sound matched the actual order.	
Arranged the bottles from actual highest to lowest sound and recorded the actual order by coloring in the level of the water in the bottles.	
Compared the accuracy of the predicted order of the bottles to the actual order of the bottles from highest to lowest sound.	

Comments:

Total	/28

Predicting

1. Tap the bottle with the most water.
2. Tap the bottle with the least water.
3. Think how each bottle will sound. Make a prediction.
4. Color in the bottles from highest sound to lowest sound.

4. Now test all five bottles. Color in the bottles in order from the highest sound to lowest sound.

5. Are the bottles in the same order as your prediction?

MAKING MODELS

Materials

Making Models Assessment Sheet
Scissors, glue, tape, crayons
craft materials: pipe cleaners, yarn, string, ribbon, cloth, tissue paper, construction paper

Procedure

Review plant parts by drawing or showing pictures of plants and discussing the following plant parts: leaf, stem, flower, seeds, fruit, and root. Then, ask the students to use the craft materials to make a model of these plant parts: leaf, stem, flower, seeds, fruit, and root. Instruct the students to glue or tape the craft materials on the assessment sheet to represent the plant. Tell the students to label each plant part by writing the name next to the part represented on the model plant. Ask each student how his/her model is like a real plant.

Making Models Rubric

4 points	exceptionally accurate/reasonable, complete, detailed
3 points	adequately accurate/reasonable, complete, detailed
2 points	somewhat accurate/reasonable, complete, detailed
1 point	not accurate/reasonable, complete, or detailed
0 points	not attempted

Scoring Criteria – Making Models	Points 0-4
Followed directions using appropriate safety precautions to complete the performance assessment.	
Made a model of the following plant parts: leaf, stem, flower, seeds, fruit, and root.	
Labeled each plant part correctly as: leaf, stem, flower, seeds, fruit and root.	
Compared the model of the plant parts to real plant parts to tell how the model is like the real plant parts.	
Total	/16

Comments:

Making Models

1. Make a model of a plant from the materials.
2. Tape your plant model below.
3. Label the parts of the plant.

leaf	stem	root	flower	fruit	seeds

4. Tell how your model is like the real plant. 

Level 2 Student Name: ______________________ Date: ______________

OBSERVING

Materials

Level 2 Observing Assessment Sheet
crayons, markers, or colored pencils
5 nuts of the same kind per student (suggested nuts: Brazil nut, walnut, acorn, pecan, peanut, almond, filbert)

Procedure

Give the student 5 nuts of the same kind. Have the student select a nut from the group of five nuts. Ask the student to draw a picture of the nut in the box. Then instruct the student observe this nut using the senses of sight and touch and to describe its color, shape, size, texture, and unique features. Record the student's responses. (Note: Students with adequate writing skills can complete the assessment sheet on their own.) Then mix the selected nut with the other four nuts. Ask the student to pick out the selected nut among the other nuts and draw a picture of the nut that shows the features that helped to find it. Then have the student describe those features that helped in picking out the selected nut among the other nuts.

Observing Rubric

4 points	exceptionally accurate/reasonable, complete, detailed
3 points	adequately accurate/reasonable, complete, detailed
2 points	somewhat accurate/reasonable, complete, detailed
1 point	not accurate/reasonable, complete, or detailed
0 points	not attempted

Scoring Criteria- Observing	Points 0-4
Followed directions using safe procedures to complete the performance assessment.	
Selected a nut and drew a picture of the nut in the box.	
Described the color, shape, size, feel, and special features of the nut.	
Found the nut after mixing it with the other nuts and drew a picture of the nut identifying the special features of the nut.	
Described features that helped to find the nut among the other nuts.	
Total	/20

Comments:

Observing

1. Choose one nut. Observe that nut with your eyes and hands.

2. Draw a picture of the nut in this box. →

3. Describe the nut.

Color	
Shape	
Size	
Feel	
Special Features	

4. Mix your nut with other nuts.

5. Find your nut and draw a picture of it in this box. →

6. What features helped you find your nut?

Level 2 Student Name: ______________________________ Date: ___________

COMMUNICATING

Materials

Level 2 Communicating Assessment Sheet
3 apples (1 whole apple 1 apple cut horizontally, and 1 apple cut vertically)
Lemon juice (put on cut surfaces of the apple keep the cut apples fresh)
Crayons, markers, or colored pencils

Procedure

Give the student a whole apple. Have the student draw and color the whole apple. Then have the student observe the vertically cut apple and draw a diagram to show what the apple looks like inside. Finally, have the student observe the horizontally cut apple and draw a diagram of this cross section. Ask the student to describe an apple from the inside to the outside. Record student responses on the assessment sheet. (Students with adequate writing skills can complete it on their own.)

Communicating Rubric

4 points exceptionally accurate/reasonable, complete, detailed
3 points adequately accurate/reasonable, complete, detailed
2 points somewhat accurate/reasonable, complete, detailed
1 point not accurate/reasonable, complete, or detailed
0 points not attempted

Scoring Criteria- Communicating	Points 0-4
Followed directions using safe procedures to complete the performance assessment.	
Drew and colored a picture of the whole apple.	
Drew and colored a picture of the inside of the apple cut vertically from top to bottom.	
Drew and colored a picture of the inside of the apple cut horizontally across the middle.	
Described the apple from the inside to the outside.	
Total	/20

Comments:

Communicating

1. Draw and color a picture of the whole apple.

2. Draw and color a picture of the inside of the apple from top to bottom.

3. Draw and color a picture of the inside of the apple cut across the middle.

4. Describe the apple from the inside to the outside.

____

____

____

Level 2 Student Name: ______________________________ Date: ______________

ESTIMATING

Materials

Level 2 Estimating Assessment Sheet
1 centimeter square piece of paper or cardboard

Procedure

Explain that the square is a centimeter long on each side and that it can be used to estimate surface area by placing it on the surface of a figure and estimating how many square centimeters would cover the surface. Ask the student to use the square to estimate the number of square centimeters that would cover each figure on the assessment sheet, and to write each estimate in the circle next to the figure. (Note: Depending on the amount of experience your student may have with finding surface area, you may want to suggest the procedure of tracing the square centimeter until each figure is filled with traced square centimeters to help estimate.)

Estimating Rubric

4 points exceptionally accurate/reasonable, complete, detailed
3 points adequately accurate/reasonable, complete, detailed
2 points somewhat accurate/reasonable, complete, detailed
1 point not accurate/reasonable, complete, or detailed
0 points not attempted

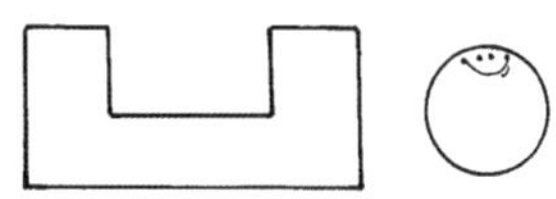

Scoring Criteria- Estimating	Points 0-4
Followed directions using safe procedures to complete the performance assessment.	
Used the square centimeter to estimate the surface area of each figure.	
Wrote an estimate of the surface area for each figure.	
Explained the strategy used to estimate.	
Total	/20

Comments:

Estimating

1. Use the square to estimate the number of square centimeters in each of the shapes below.

2. Write your estimate in the circle next to the shape.

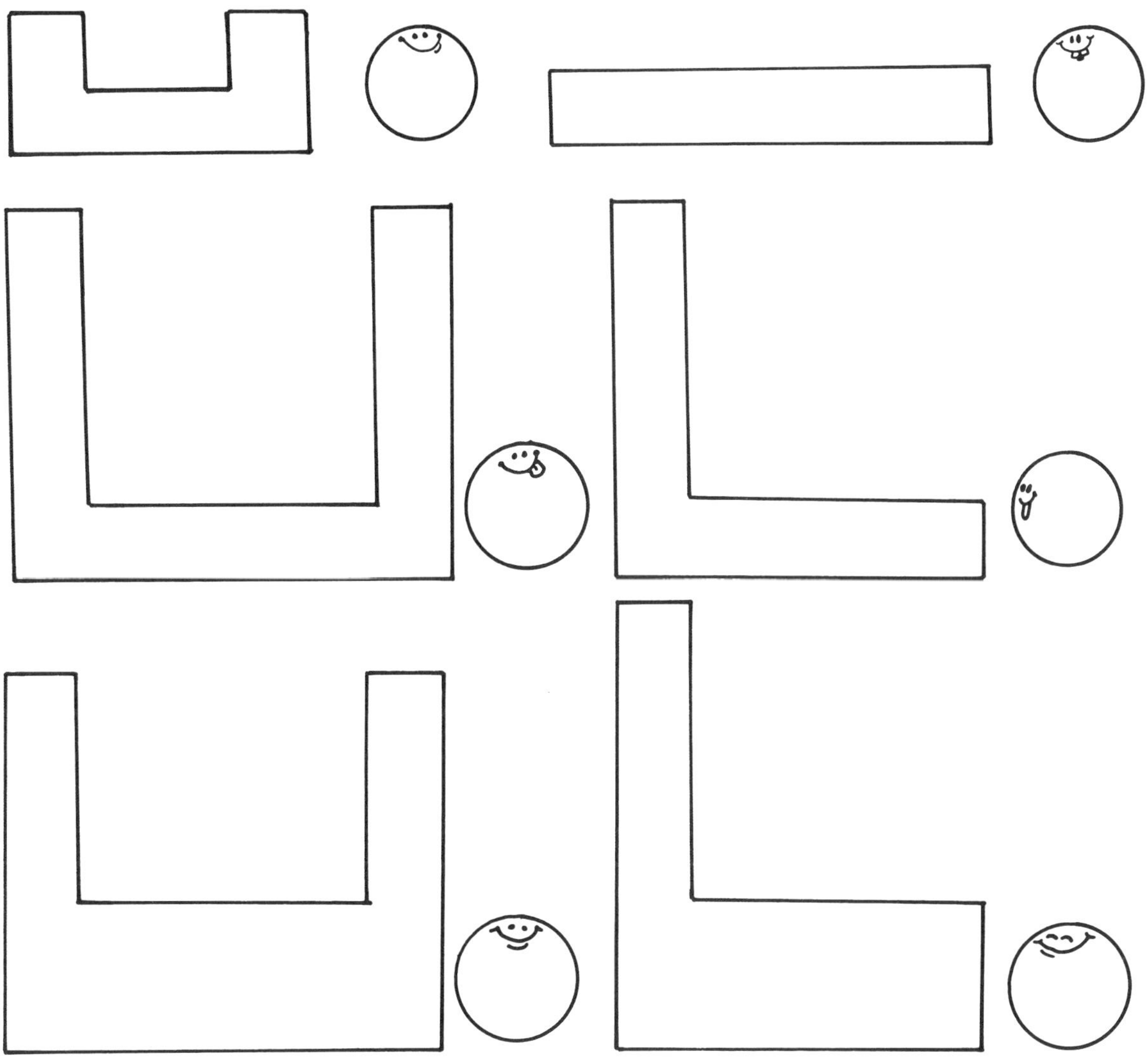

3. Explain how you estimated.

____

____

Level 2 Student Name ______________________________ Date: ______________

MEASURING

Materials

Level 2 Measuring Assessment Sheet
6 containers of various sizes labeled A, B, C, D, E
100-ml measuring cup
Water

Procedure

Ask students to point to the container they estimate would hold the smallest amount of water. Then ask them to point to the container they estimate would hold the largest amount of water. Instruct students to order the containers from the one that will hold the smallest amount of water to the one that will hold the largest amount of water and write the letter of the containers in order from smallest amount of water to largest amount of water on the *Measuring Assessment Sheet.* Then have students use the measure cup to fill each container with water and record the number of cups required to fill each cup. Ask students which container held the fewest measuring cups of water. Instruct students to order the containers from the one that held the least number of measuring cups of water to the one that held the most measuring cups of water and record the letter of the containers in order. Have students compare their estimate to their actual results when the measuring cup was used to determine how much water each one held. Ask students to record which containers surprised them and why they the container was surprising.

Measuring Rubric

4 points	exceptionally accurate/reasonable, complete, detailed
3 points	adequately accurate/reasonable, complete, detailed
2 points	somewhat accurate/reasonable, complete, detailed
1 point	not accurate/reasonable, complete, or detailed
0 points	not attempted

Scoring Criteria- Measuring	Points 0-4
Followed directions using safe procedures to complete the performance assessment.	
Recorded the estimate of the order of the containers from smallest to largest.	
Recorded the number of measuring cups required to fill each container with water.	
Recorded the arrangement of the cups from smallest to largest based on the number of measuring cups required to fill each container.	
Compared the estimate to the actual number of measuring cups of water required to fill each container and indicated which containers were surprising.	
Total	/20

Comments:

Measuring

1. Estimate the order of containers from smallest to largest. Write the letter of each container in the boxes.

smallest to → largest

2. Use the measuring cup to fill each container with water. Write the number of cups of water you used to fill each container.

A	B	C	D	E
cups	cups	cups	cups	cups

3. Arrange the containers from smallest to largest. Write the letter of each container in order.

smallest to → largest

4. Compare your estimate to the number of cups each container held. Which containers surprised you?

____

____

Level 2 Student Name: ______________________________ Date: ______________

COLLECTING DATA

Materials

Level 2 Collecting Data Assessment Sheet
30 buttons or other objects of the same size but different colors
(use different number of buttons for each color; e.g., 8 white, 6 black, 4 red, 2 blue, 7 pink, 3 yellow)
Crayons, markers, or colored pencils (in colors matching colors of the buttons or objects)
Note: Buttons/objects should be the same size or smaller than the squares in the graph paper on the assessment sheet.

Procedure

Give the student 30 buttons or other small objects (same size or smaller than the squares on the graph paper). Have the student sort the buttons or objects into piles according to a rule, such as color. Then have the student place the objects that are in the same pile in the squares on the graph beginning at the bottom of the graph and going up consecutive squares in a column. Ask the student to trace and color the button or object in each column. Then ask the student to count the squares that they traced in each column and write the number in the circle below each column. Have the student write three true statements about the graph, e.g.; *There are more red buttons than yellow buttons.*

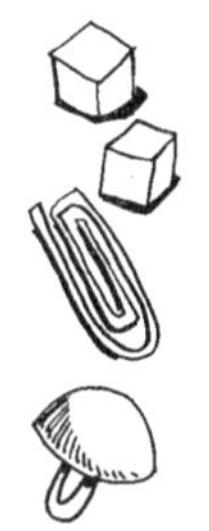

Collecting Data Rubric

4 points	exceptionally accurate/reasonable, complete, detailed
3 points	adequately accurate/reasonable, complete, detailed
2 points	somewhat accurate/reasonable, complete, detailed
1 point	not accurate/reasonable, complete, or detailed
0 points	not attempted

Scoring Criteria-Collecting Data	Points 0-4
Followed directions using safe procedures to complete the performance assessment.	
Used a rule to sort the objects into groups and recorded the rule used to sort the objects.	
Placed the objects on the graph by lining up the sorted objects in each pile in a column beginning at the bottom of the graph and then traced each object.	
Labeled the graph to indicate the number of objects in each column.	
Analyzed the graph by writing three true statements about the graph.	
Total	/20

Comments:

Collecting Data

1. Sort the objects into groups by a rule.
   Rule: ______________________________
2. Place the same kind of object in a column on the grid.
3. Color in the spaces to make a graph. Label.

Level 2 Student Name ______________________ Date: ____________

CLASSIFYING

Materials

Level 2 Classifying Assessment Sheet
penny
nickel
dime
quarter
(Note: Select some coins with the same date.)

Procedure

Give the student a penny, nickel, dime, and quarter. Ask the student to sort the coins into two groups and place them in the two loops. Have the student trace around each coin. Then have the student name each group and tell how the coins in each group are different. (You can record responses or have students with adequate writing skills complete the recording sheet with their own responses.) Then ask the student to sort the coins into two groups in a different way. Have the student trace around each coin and name each group. Finally, have the student explain how the coins in each group are alike; e.g. shape, round, circles, money, coins, all have dates, etc.

Classifying Rubric

4 points exceptionally accurate/reasonable, complete, detailed
3 points adequately accurate/reasonable, complete, detailed
2 points somewhat accurate/reasonable, complete, detailed
1 point not accurate/reasonable, complete, or detailed
0 points not attempted

Scoring Criteria- Classifying	Points 0-4
Followed directions using safe procedures to complete the performance assessment.	
Sorted the coins into two groups, traced the coins, and named the groups.	
Described how the coins in each group are different.	
Sorted the coins into two groups in a different way, traced the coins, and named the groups.	
Described how the coins in each group are alike.	
Total	/20

Comments:

Classifying

1. Sort the coins into two groups below.

Name: ____________________ Name: ____________________

2. Trace around each coin. Name each group.
3. How are the coins in each group different?

____

____

4. Sort the coins into two groups a different way.

Name: ____________________ Name: ____________________

5. Trace around each coin. Name each group.
6. How are the coins in each group alike?

____

____

Level 2 Student Name: ______________________________ Date: __________

INFERRING

Materials

Level 2 Inferring Assessment Sheet
crayons, markers, or colored pencils

Procedure

Ask the student to observe the pattern in each row and infer what picture would come next. Have the student draw and color the next picture in the box at the end of rows A, B, C and D. Then have the student create, draw, and color a pattern in the last row.

Inferring Rubric

4 points	exceptionally accurate/reasonable, complete, detailed
3 points	adequately accurate/reasonable, complete, detailed
2 points	somewhat accurate/reasonable, complete, detailed
1 point	not accurate/reasonable, complete, or detailed
0 points	not attempted

Scoring Criteria- Inferring	Points 0-4
Followed directions using safe procedures to complete the performance assessment.	
Inferred which picture would come next in the first four rows.	
Drew and colored the picture that would come next in the first four rows.	
Created a pattern for the last row.	
Drew and colored a pattern in the last row.	
Total	/20

Comments:

Inferring

1. Look at the pattern in each row.
2. Draw what comes next.
3. Make your own pattern at the bottom.

Row A

Row B

Row C

Row D

Your Row

Level 2 Student Name: ______________________ Date: ____________

PREDICTING

Materials

Level 2 Predicting Assessment Sheet
magnet
aluminum can, paper clip, plastic spoon, cardboard, steel nail, penny screw, sponge, straw, steel can, aluminum foil, wood

Procedure

Ask the students to predict which objects will be attracted to a magnet by circling the objects on the assessment sheet. Then give the student a magnet, aluminum can, paper clip, plastic spoon, cardboard, steel nail, penny screw, sponge, straw, steel can, aluminum foil, and wood. Have students test their predictions by placing the magnet on each object. Tell them to put an X on each object that does not stick to the magnet.

Predicting Rubric

4 points	*exceptionally accurate/reasonable, complete, detailed*
3 points	*adequately accurate/reasonable, complete, detailed*
2 points	*somewhat accurate/reasonable, complete, detailed*
1 point	*not accurate/reasonable, complete, or detailed*
0 points	*not attempted*

Scoring Criteria- Predicting	Points 0-4
Followed directions using safe procedures to complete the performance assessment.	
Predicted which objects will be attracted to a magnet and circled the objects on the assessment sheet.	
Tested each object with a magnet to observe which objects are attracted and which objects are not attracted to a magnet.	
Put an X on each object on the assessment sheet that does not stick to a magnet.	
Described how all objects that stick to a magnet are alike.	
Total	/20

Comments:

Predicting

1. Circle the objects you think will be attracted to a magnet.
2. Test each object with a magnet.
3. Put an X on each object that does not stick.

 aluminum can	 paper clip	 plastic spoon	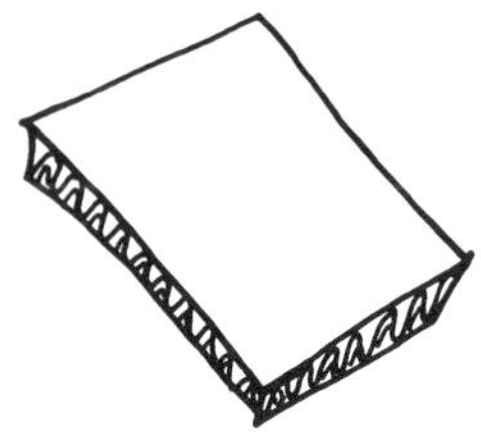 cardboard
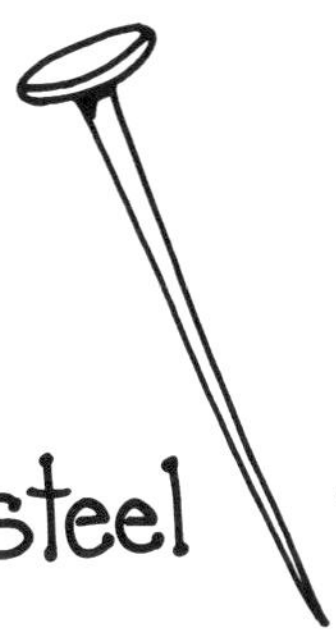 steel nail	 penny	 screw	 sponge
 straw	 steel can	 aluminum foil	 wood

4. How are all objects that stick to a magnet alike?

Level 2 Student Name: ______________________________ Date: ____________________

MAKING MODELS

Materials

Level 2 Making Models Assessment Sheet
Scissors
Paper clip
Winged seed models

Procedure

Have the student cut out the model of a winged seed on the solid lines. Ask the student to fold the paper on the dotted lines and attach a paper clip to the bottom. Then have the student drop the paper model and observe its motion. Ask the student to explain how this model shows the way the wind disperses seeds.

Making Models Rubric

4 points	exceptionally accurate/reasonable, complete, detailed
3 points	adequately accurate/reasonable, complete, detailed
2 points	somewhat accurate/reasonable, complete, detailed
1 point	not accurate/reasonable, complete, or detailed
0 points	not attempted

Scoring Criteria- Making Models	Points 0-4
Followed directions using safe procedures to complete the performance assessment.	
Cut out, folded, and attached a paper clip to the bottom of the winged seed model.	
Dropped the winged seed model and observed how it moves.	
Explained how the winged seed model shows how wind carries seeds.	
Described something that could be done to the model to make it more like a winged seed model.	
Total	/ 20

Comments:

Making Models

1. Cut out the winged seed model on the solid lines.
2. Fold the model on the dotted lines.

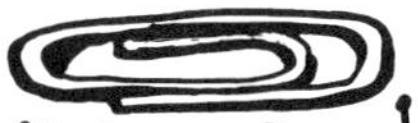

3. Attach a paper clip to the bottom of the model for balance.
4. Drop the seed model and observe how it moves.

5. Explain how your model shows how wind carries seeds.

____

____

____

____

6. What else could you do to your model to make it more like a winged seed?

____

____

____

____

Winged Seed Models - Level 2

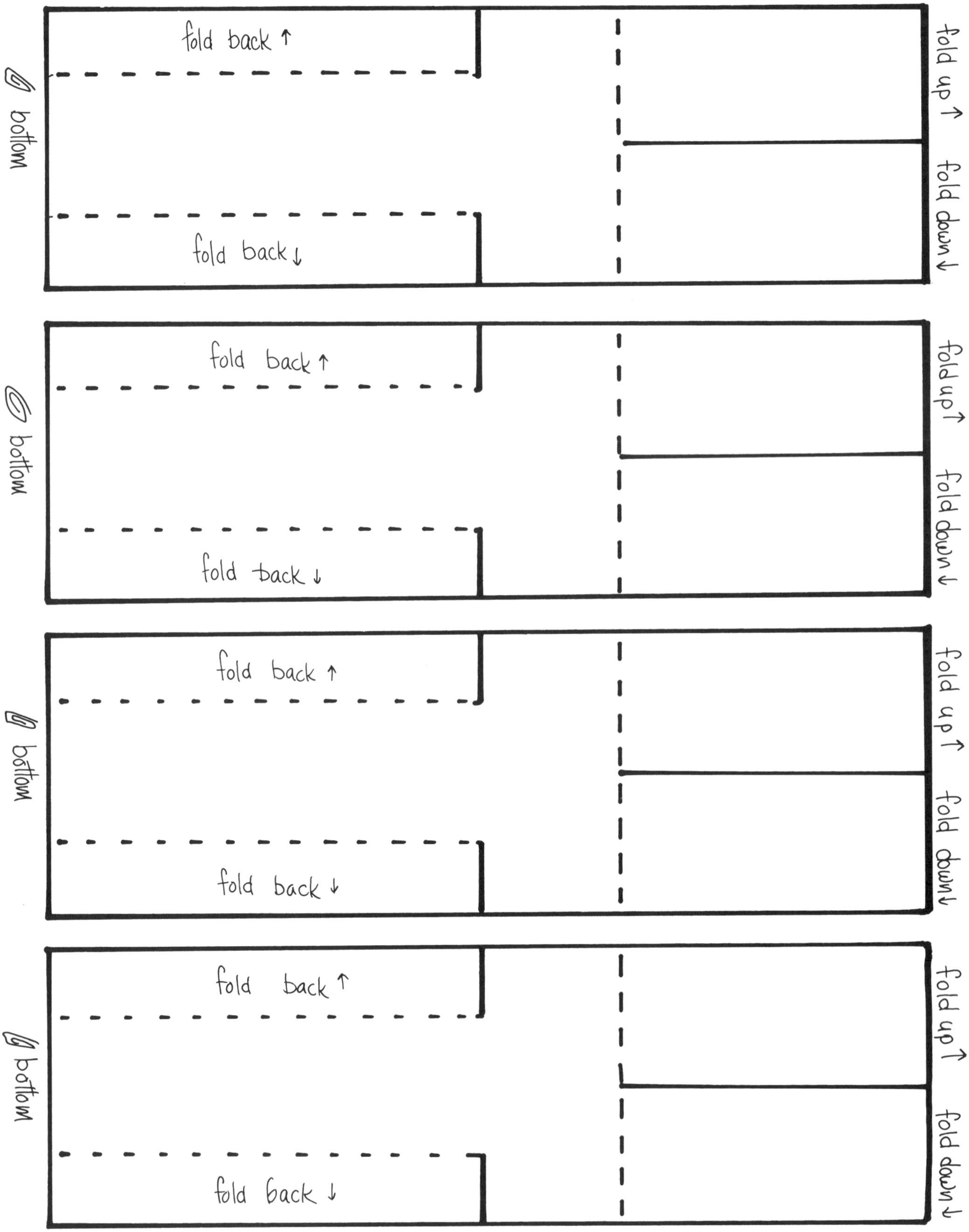

Level 3 Student Name: ______________________ Date: __________

OBSERVING

Materials

Level 3 Observing Assessment Sheet
2 bar magnets
2 paper clips

Procedure

Ask students why accurate observations are important in science. Explain that students will have an opportunity to demonstrate that they can make and describe accurate observations of magnetism on this performance assessment. Instruct students to touch one end of a bar magnet to a paper clip and describe their observations using their senses of sight and touch. Have students touch one end of a paper clip to several places along the magnet and describe what happens. Next, have students slide the N end of a bar magnet toward the S end of another bar magnet. Then ask students to observe and describe what happens when the N end of one bar magnet is slid toward the N end of another bar magnet. Challenge students to suspend two paper clips side by side from the lower end of a bar magnet and describe their observations. Finally, challenge students to make several attempts to put a paper clip between two magnets with N and S ends together and slowly pull the magnets apart until the paper clip disconnects.

Observing Rubric

4 points	exceptionally accurate/reasonable, complete, detailed
3 points	adequately accurate/reasonable, complete, detailed
2 points	somewhat accurate/reasonable, complete, detailed
1 point	not accurate/reasonable, complete, or detailed
0 points	not attempted

Scoring Criteria- Observing	Points 0-4
Followed directions using safe procedures to complete the performance assessment.	
Described observations when a paper clip is touched by one end of a bar magnet.	
Described observations when a paper clip is touched to several places along a bar magnet.	
Described observations when the N end of a bar magnet is slid toward the S end of another bar magnet.	
Described observations when the N end of a bar magnet is slid toward the N end of another bar magnet.	
Described observations when 2 paper clips are suspended side by side from the lower end of a bar magnet.	
Described observations when a paper clip is put between 2 magnets with the N and S ends together and slowly pulled apart until the paper clip disconnects.	
Total	/28

Comments:

Observing

Describe what happens each time.

1. Touch one end of a bar magnet to a paper clip. 	**4.** Slide the N end of a magnet toward the N end of another. 
2. Touch one end of a paper clip to several places along the magnet. 	**5.** Suspend 2 paper clips from lower end of the magnet side by side. 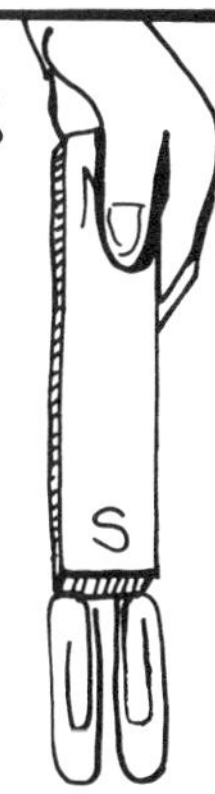
3. Slide the N end of one magnet toward the S end of another. 	**6.** Put a paper clip between 2 magnets with N and S ends together. Slowly pull the magnets apart until the paper clip disconnects. Try several times. 

Level 3 Student Name________________________ Date__________

COMMUNICATING

Materials

Level 3 Communicating Assessment Sheet
1 peanut in the shell per student
crayons, markers, or colored pencils

Procedure

Ask a student to describe a favorite toy in words. Then ask students if a drawing would be useful to communicate effectively so that everyone could exactly what the toy looks like. Tell students that they will complete the performance assessment by communicating using drawings and words. Give the student a peanut in the shell. Have the student observe it closely, trace the peanut and draw the shell's pattern. Then ask the student to describe the shell in words. Instruct the student to gently break the shell in order to separate the shell into two halves. Ask the student to draw what the inside of the peanut shell looks like and describe it in words. Finally, have the student gently roll a peanut between two fingers to break apart the peanut in order to separate it into two halves. Ask the student to draw the inside of the peanut and describe the inside in words. Then ask the student to describe the most surprising part of this peanut.

Communicating Rubric

4 points	exceptionally accurate/reasonable, complete, detailed
3 points	adequately accurate/reasonable, complete, detailed
2 points	somewhat accurate/reasonable, complete, detailed
1 point	not accurate/reasonable, complete, or detailed
0 points	not attempted

Scoring Criteria-Communicating	Points 0-4
Followed directions using safe procedures to complete the performance assessment.	
Traced peanut, drew shell's pattern, and described the shell in words.	
Drew the inside of the peanut shell and described the inside of the shell in words.	
Drew the inside of the peanut and described the inside in words.	
Described the part of the peanut that was the most surprising.	
Total	/20

Comments:

Communicating

1. Examine the peanut closely.

Trace the peanut and draw in the shell's pattern.

Describe the shell in words.

2. Gently break the shell and separate the two halves.

Draw the inside of the peanut shell.

Describe the inside of the shell in words.

3. Gently break apart one peanut and separate the two halves.

Draw the inside of the peanut

Describe the inside in words.

4. What part of this peanut was the most surprising?

Level 3 Student Name: ______________________________ Date: ______________

ESTIMATING

Materials

Level 3 Estimating Assessment Sheet
6 different sized and shaped containers filled with objects
(Label 5 of the Containers #1, #2, #3, #4, and #5).

Procedure

Show students a container of objects. Ask students to estimate the number of objects in the unlabeled container and put the estimate on a Post-It® with their name. Line the estimates up in order from the lowest to highest estimate. Then pour out the objects in piles and select students to count a pile of objects. Add the numbers up to get the total number of objects. Identify the estimates that were the most accurate and ask those students to describe the technique they used to make their estimate. Then challenge the students to use those techniques to estimate the number of objects in 5 containers. After estimating, ask students to count to determine the actual number of objects. Challenge students to use estimation techniques to make their estimates more accurate as they estimate and count the objects in each container.

Estimating Rubric

4 points exceptionally accurate/reasonable, complete, detailed
3 points adequately accurate/reasonable, complete, detailed
2 points somewhat accurate/reasonable, complete, detailed
1 point not accurate/reasonable, complete, or detailed
0 points not attempted

Scoring Criteria- Estimating	Points 0-4
Followed directions using safe procedures to complete the performance assessment.	
Estimated the number of objects in Container #1 and recorded the actual count.	
Estimated the number of objects in Container #2 and recorded the actual count.	
Estimated the number of objects in Container #3 and recorded the actual count.	
Estimated the number of objects in Container #4 and recorded the actual count.	
Described the procedure used to estimate.	
Estimated the number of objects in Container #5 and recorded the actual count.	
Described the procedure used to estimate.	
Total	/28

Comments:

Estimating

1. Estimate and then count the number of objects each container holds.

Container	Estimate	Actual Count
Container # 1		
Container # 2		
Container # 3		
Container # 4		

2. Describe what you did to estimate.

____

____

____

3. Use your results to help you estimate how many objects are in container #5.

Estimate: ☐ Actual Count: ☐

4. Describe what you did to estimate.

____

____

____

____

Level 3 Student Name: ______________________________ Date: __________

MEASURING

Materials

Level 3 Measuring Assessment Sheet
Collection of 9 objects labeled A, B, C. D, E. F, G, H, and I
Gram weights
Balance

Procedure

Ask students to review the steps involved in using a balance to mass objects. Display the labeled objects, gram weights, and balance. Explain that they will estimate and determine the mass of the 9 objects and record this information on the measuring performance assessment. Demonstrate how to hold an object in one hand and adjust the number of gram weights in the other hand to estimate the mass of the object. Have students record the name of the object, their estimate of the object's mass, and then use the balance to find the actual mass of each object. Then ask students to list the letter on each object in the boxes at the bottom of the sheet to order them from least mass to most mass.

Measuring Rubric

4 points exceptionally accurate/reasonable, complete, detailed
3 points adequately accurate/reasonable, complete, detailed
2 points somewhat accurate/reasonable, complete, detailed
1 point not accurate/reasonable, complete, or detailed
0 points not attempted

Scoring Criteria- Measuring	Points 0-4
Followed directions using safe procedures to complete the assessment.	
Recorded the name of object A, the estimated mass, and actual mass.	
Recorded the name of object B, the estimated mass, and actual mass.	
Recorded the name of object C, the estimated mass, and actual mass.	
Recorded the name of object D, the estimated mass, and actual mass.	
Recorded the name of object E, the estimated mass, and actual mass.	
Recorded the name of object F, the estimated mass, and actual mass.	
Recorded the name of object G, the estimated mass, and actual mass.	
Recorded the name of object H, the estimated mass, and actual mass.	
Recorded the name of object I, the estimated mass, and actual mass.	
Recorded the letter of each object in order from least mass to most mass.	
Total	/ 44

Comments:

Measuring

1. Gather objects to measure. Hold an object in one hand and gram weights in the other.
2. Estimate the mass of each object in grams.
3. Use a balance to find the actual mass. Place an object on one side and grams on the other until they balance. Record.

Object	Estimated Mass	Actual Mass
A.	grams	grams
B.	grams	grams
C.	grams	grams
D.	grams	grams
E.	grams	grams
F.	grams	grams
G.	grams	grams
H.	grams	grams
I.	grams	grams

4. Put the objects in order from least mass to most mass.

least →								most

COLLECTING DATA

Materials

Level 3 Collecting Data Assessment Sheet
One shoe from each student

Procedure

Ask students what kinds of data (information) they have collected and what techniques they have used to the collect data. Tell them that they will have an opportunity to collect data about the kind of shoes that their classmates are wearing today on the Collecting Data Performance Assessment. Make a labels for "Girls – Slip-on," "Girls – Laces," "Girls – Buckles," "Girls – Velcro," and "Girl's Other," and "Boys – Slip-on," "Boys – Laces," "Boys – Buckles," "Boys – Velcro," and "Boys Other," Ask students to remove one of their shoes and to line the shoes up by the appropriate sign to make a concrete bar graph. Then ask students to make tally marks to represent the number of shoes of each type. Tell students to count the tally marks and record the count for each type of shoe under "Girls" and "Boys." Then have them write three true statements about the data they collected.

Collecting Data Rubric

4 points exceptionally accurate/reasonable, complete, detailed
3 points adequately accurate/reasonable, complete, detailed
2 points somewhat accurate/reasonable, complete, detailed
1 point not accurate/reasonable, complete, or detailed
0 points not attempted

Scoring Criteria- Collecting Data	Points 0-4
Followed directions using safe procedures to complete the performance assessment.	
Accurately used tally marks to recorded the number of **slip-on** shoes for girls and boys and gave the actual count.	
Accurately used tally marks to recorded the number of **lace** shoes for girls and boys and gave the actual count.	
Accurately used tally marks to recorded the number of **buckle** shoes for girls and boys and gave the actual count.	
Accurately used tally marks to recorded the number of **Velcro** shoes for girls and boys and gave the actual count.	
Accurately used tally marks to recorded the number of **other** shoes for girls and boys and gave the actual count.	
Wrote three true statements about the data collected.	
Total	/28

Comments:

Collecting Data

1. Observe the types of shoes your classmates are wearing today.

2. Make a tally mark for each type you see. Count them and compare.

Type of Shoe	Girls		Boys	
	Tally	Count	Tally	Count
Slip-on				
Laces				
Buckles				
Velcro				
Other				

3. Write three true statements about your data.

① ____

② ____

③ ____

 Student name________________________ Date____________

CLASSIFYING

Materials

Level 3 Classifying Assessment Sheet
Crayons
Assortment of seeds in small resealable plastic bag
(15 bean soup mix can be used to make the bags of assorted seeds)

Procedure

Explain that in order to classify, students need to make careful observations to determine the similarities and differences in objects. Give each student a bag of assorted seeds to observe. Then ask students to think of one way the seeds are different and divide the seeds into two groups based on this difference. Have the students place the seeds in the boxes on the assessment sheet, trace around the seeds, and color them. Tell them to name each group and write that name in the box. Then ask students to describe how the seeds in each group are different. Challenge students to think of another way they can sort the seeds into two groups. Have them place the seeds in the second set of boxes on the assessment sheet, trace the seeds in the box, color them, and name each group. Then ask students to explain how the seeds in each group are the same.

Classifying Rubric

4 points	exceptionally accurate/reasonable, complete, detailed
3 points	adequately accurate/reasonable, complete, detailed
2 points	somewhat accurate/reasonable, complete, detailed
1 point	not accurate/reasonable, complete, or detailed
0 Points	not attempted

Scoring Criteria- Classifying	Points 0-4
Followed directions using appropriate safety precautions to complete the performance assessment.	
Sorted the seeds into two groups, traced around the seeds, colored the seeds, and named each group.	
Explained how the seeds in each group are different.	
Sorted the seeds into two groups a different way, traced around the seeds, colored the seeds, and named each group.	
Explained how the seeds in each group are the same.	
Total	/20

Comments:

Classifying

1. Sort the seeds into two groups in the boxes below.
2. Trace around the seeds and color them. Name each group.

Name:	Name:

3. How are the seeds in each group different?

4. Sort the seeds in a different way in the boxes below.
5. Trace around the seeds and color them. Name each group.

Name:	Name:

6. How are the seeds in each group the same?

Level 3 Student Name: ______________________ Date: ____________

INFERRING

Materials

Level 3 Inferring Assessment Sheet
6 squares of fabric approximately 5 cm square (2 inches square).
Crayons

Procedure

Advanced Preparation: Make one "Mystery Rubbing" for each student of one of the pieces of fabric by placing a sheet of paper over the fabric and gently rubbing with a dark colored crayon. Show students six pieces of fabric and ask them to describe each one using their senses of sight and touch. Explain that these observations are direct evidence gathered through the senses. Then place a sheet of paper over each piece of fabric and demonstrate how to use a crayon to do a rubbing of the fabric. Show the students one of the rubbings and ask them to infer which piece of fabric was used to make the rubbing. Tell the students that inferences are explanations or interpretations based on observations. Then have students make rubbings by placing each fabric under the squares on the assessment sheet. Have them circle the rubbing that matches the "Mystery Rubbing" and then explain what helped them identify the fabric that was used to make the "Mystery Rubbing."

Inferring Rubric

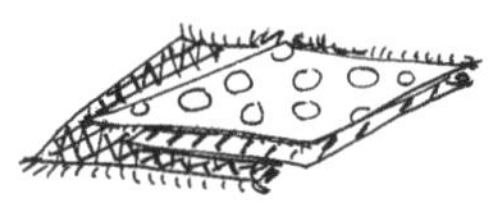

4 points — exceptionally accurate/reasonable, complete, detailed
3 points — accurate/reasonable, complete, detailed
2 points — slightly accurate/reasonable, complete, detailed
1 point — not accurate/reasonable, complete, or detailed
0 points — not attempted

Scoring Criteria- Inferring	Points 0-4
Followed directions using safe procedures to complete the performance assessment.	
Used a crayon to make a rubbing of one of the squares of fabric in Square A on the assessment sheet.	
Used a crayon to make a rubbing of one of the squares of fabric in Square B on the assessment sheet.	
Used a crayon to make a rubbing of one of the squares of fabric in Square C on the assessment sheet.	
Used a crayon to make a rubbing of one of the squares of fabric in Square D on the assessment sheet.	
Used a crayon to make a rubbing of one of the squares of fabric in Square E on the assessment sheet.	
Used a crayon to make a rubbing of one of the squares of fabric in Square F on the assessment sheet.	
Circled the rubbing that matches the "Mystery Rubbing" and explained what helped to identify the fabric.	

Comments:

Total	/36

Inferring

1. Find the Mystery Rubbing. Make a crayon rubbing by placing each fabric under a square. Then, gently rub a crayon across the top of the square.

A.

B.

C.

D.

E.

F.

2. Mark the rubbing of the fabric that was used to make the Mystery Rubbing.

3. Explain what helped you identify the fabric that was used to make the Mystery Rubbing.

____

____

PREDICTING

Materials

Predicting Level 3 Assessment Sheet
White Plastic Desert Plate, Crayons, Water
3 pieces of candy (i.e. gumballs, M&M's®, or Gobstoppers®)
Waste container for used water and candies (plastic deli container)

Procedure

Ask students to describe what happens when a solid and a liquid are mixed together. Explain that they will place solids in a liquid and observe how they interact. Instruct students to place a candy in the middle of a white plastic desert plate partially filled with water. *Note: Place just enough water in the plate so that the water doesn't go over the top when the candy is added to the water.* Have students observe, describe, and draw the pattern resulting from the interaction of the candy and the water. Then have students dispose of the water and candy in the waste container. Explain to students that predictions are based on prior experiences and data that have been collected. Ask them to predict the pattern that two candies will make when they interact with water by both describing and drawing. Then have students test their predictions by placing two candies in a plate of clean water, drawing and describing the results. Next, have them compare their results to their prediction. Explain that predictions aren't "right" or "wrong," but that some predictions are more accurate than other predictions and they should try to make accurate predictions based on prior experiences and data collected. Have students dispose of the candies and water in the waste container.

Predicting Rubric

4 points	exceptionally accurate/reasonable, complete, detailed
3 points	adequately accurate/reasonable, complete, detailed
2 points	somewhat accurate/reasonable, complete, detailed
1 point	not accurate/reasonable, complete, or detailed
0 points	not attempted

Scoring Criteria- Predicting	Points 0-4
Followed directions using appropriate safety precautions to complete the performance assessment.	
Drew and described what happens when a candy is placed in a dish of water.	
Predicted, described, and drew what **will happen** when two candies are placed in a dish of water.	
Drew and described what happens when two candies **are** placed in a dish of water.	
Compared the predicted results with the actual results of placing two candies in a dish of water.	
Total	/20

Comments:

Predicting

1. Place one candy in the middle of a plate of water. Observe.
2. Describe and draw what you see.
3. Predict: What will the pattern look like if you place two different colors of candies in a fresh dish of water. Draw and describe.
4. Test: Place the candies in water and observe. Draw and describe what happens.
5. How do your results compare to your prediction?

Level 3 Student Name: ______________________ Date: ____________

MAKING MODELS

Materials

Level 3 Making Models Assessment Sheet
Glue
Cotton

Procedure

Assess prior knowledge by asking students to describe or draw clouds they have observed. Ask students what type of weather they observed with each type of cloud. Explain to students that clouds are water – either small liquid water drops or tiny pieces of ice. Tell them that meteorologists classify clouds by their height and shape (flat or fluffy). Ask students to demonstrate how to pull and stretch cotton to match the shape of stratus, cumulus, cirrus, and cumulonimbus clouds. Then instruct students to shape the cotton and then glue the cotton on the drawings of each cloud type on the assessment sheet. Ask students how this model is useful in understanding the cloud types (it shows the height and shape of clouds).

Investigating Rubric

4 points exceptionally accurate/reasonable, complete, detailed
3 points adequately accurate/reasonable, complete, detailed
2 points somewhat accurate/reasonable, complete, detailed
1 point not accurate/reasonable, complete, or detailed
0 points not attempted

Scoring Criteria- Making Models	Points 0-4
Followed directions and completed the activity.	
Shaped cotton to match the shape of stratus clouds and glued the cotton on the assessment sheet on the drawing of stratus clouds.	
Shaped cotton to match the shape of cumulus clouds and glued the cotton on the assessment sheet on the drawing of cumulus clouds	
Shaped cotton to match the shape of cirrus clouds and glued the cotton on the assessment sheet on the drawing of cirrus clouds	
Shaped cotton to match the shape of cumulonimbus clouds and glued the cotton on the assessment sheet on the drawing of cumulonimbus clouds	
Stated how the model is useful in understanding the cloud types.	

Comments: Total / 24

Making Models

1. Make models of cloud formations. Shape the cotton to resemble each type of cloud.

2. Glue the cotton to the cloud chart.

Level 3   Student Name: ______________________________ Date: ______________

INTERPRETING DATA

Materials

Level 3 Interpreting Data Assessment Sheet
Red, Yellow, and Blue Watercolors
Paint brush
Water

Procedure

Ask students what it means to interpret data. Then instruct students to use ONLY red, yellow and blue watercolors to complete the Interpreting Data performance assessment sheet. Have them mix equal amounts of red and yellow watercolors and record the color resulting from this mixture. Repeat for mixing equal amounts of yellow and blue and then equal amounts of blue and red. Then have them use these data to complete the color wheel using only red, yellow, and blue watercolors. Ask students to explain why the colors on a color wheel are arranged in this specific pattern. Then ask students to explain what would happen if unequal amounts of color are mixed.

Interpreting Data Rubric

4 points	exceptionally accurate/reasonable, complete, detailed
3 points	adequately accurate/reasonable, complete, detailed
2 points	somewhat accurate/reasonable, complete, detailed
1 point	not accurate/reasonable, complete, or detailed
0 points	not attempted

Scoring Criteria- Interpreting Data	Points 0-4
Followed directions using safe procedures to complete the performance assessment.	
Recorded the color resulting from mixing equal amounts of yellow and blue watercolors.	
Recorded the color resulting from mixing equal amounts of red and yellow watercolors.	
Recorded the color resulting from mixing equal amounts of blue and red watercolors.	
Used red, yellow, and blue watercolors to paint a color wheel red, orange, yellow, green, blue, and purple.	
Explained why a color wheel should be arranged in a specific sequence.	
Explained what happens when unequal amounts of watercolors are mixed.	
Total	/ 28

Comments:

Interpreting Data

1. Paint with <u>only</u> red, yellow and blue watercolors.
2. Mix equal amounts of paint to test. Paint your results.

Color		Color		New Color
red	+	yellow	=	
yellow	+	blue	=	
blue	+	red	=	

3. Use your data to paint a color wheel with only red, yellow and blue.

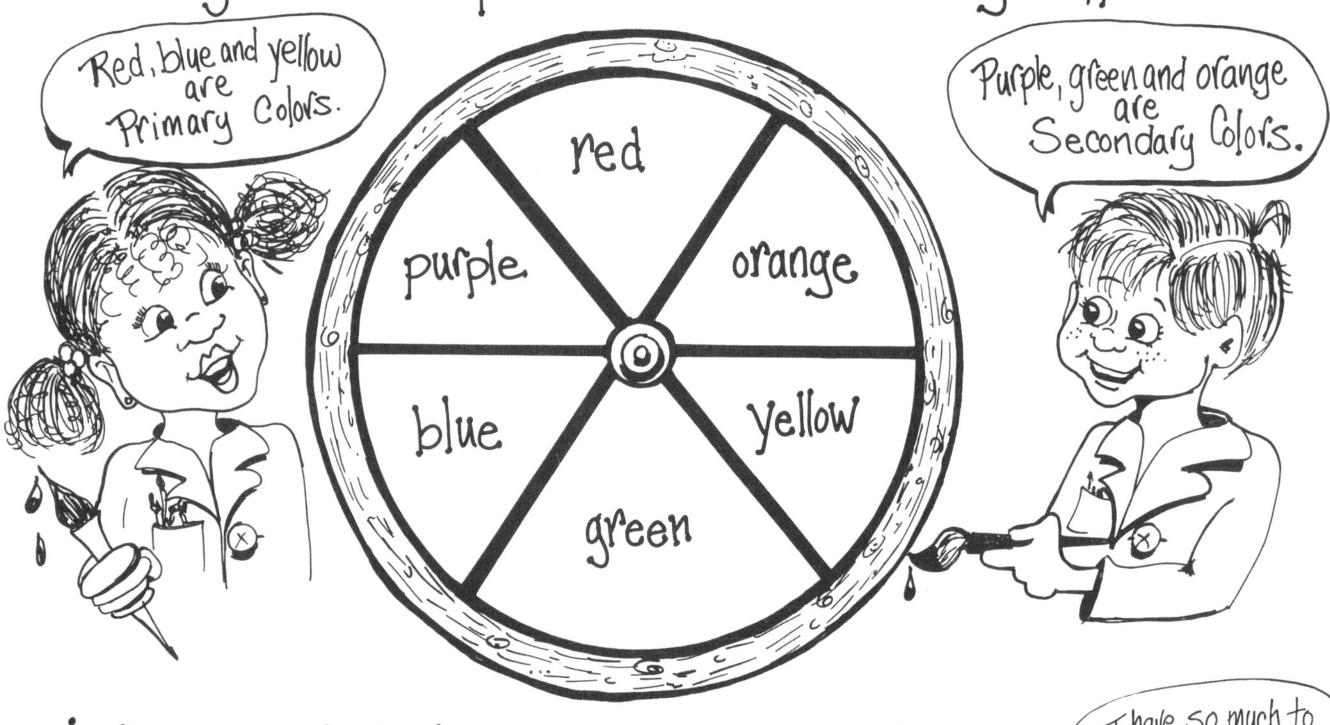

4. Why do you think the colors on a color wheel are arranged this way?

5. What happens if you mix unequal amounts of colors?

Level 3 Student Name: ______________________ Date: __________

MAKING GRAPHS

Materials

Level 3 Making Graphs Assessment Sheet
10 green bean pods per student

Procedure

Ask students how graphs help us analyze data. Explain that students will have an opportunity to make a graph called a histogram that will record the number of times that 10 green bean pods have a specific number of beans. Instruct students to open 10 green bean pods and record the number of pods that have each number of bean seeds in the pod. Then tell them to construct a histogram to show the number of times a green bean had 0 seeds, 1 seed, 2 seeds, etc. Have students use the graph to determine the most common number of seeds found in the 10 green bean pods.

Making Graphs Rubric

4 points	exceptionally accurate/reasonable, complete, detailed
3 points	adequately accurate/reasonable, complete, detailed
2 points	somewhat accurate/reasonable, complete, detailed
1 point	not accurate/reasonable, complete, or detailed
0 points	not attempted

Scoring Criteria- Making Graphs	Points 0-4
Followed directions using safe procedures to complete the performance assessment.	
Counted and recorded the number of beans in ten green bean pods.	
Constructed a histogram by coloring one square to represent each bean pod according to the number of bean seeds each bean pod had from 0 – 9 beans.	
Indicated the common number of seeds found in the ten green bean pods.	
Total	/ 16

Comments:

Making Graphs

1. Open ten green beans and count the seeds.

2. Record the number of seeds you found in each green bean.

Bean Number	1	2	3	4	5	6	7	8	9	10
Number of Seeds										

3. Color one square for each bean that had 3 seeds. Do the same for 4 seeds, 5 seeds, and so on.

Title: Seeds Found in Green Beans

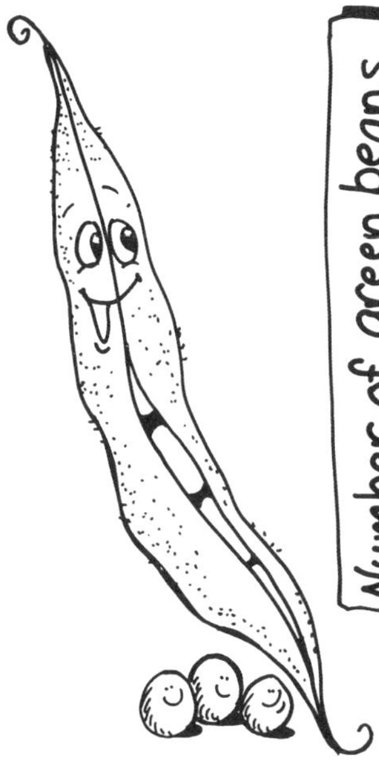

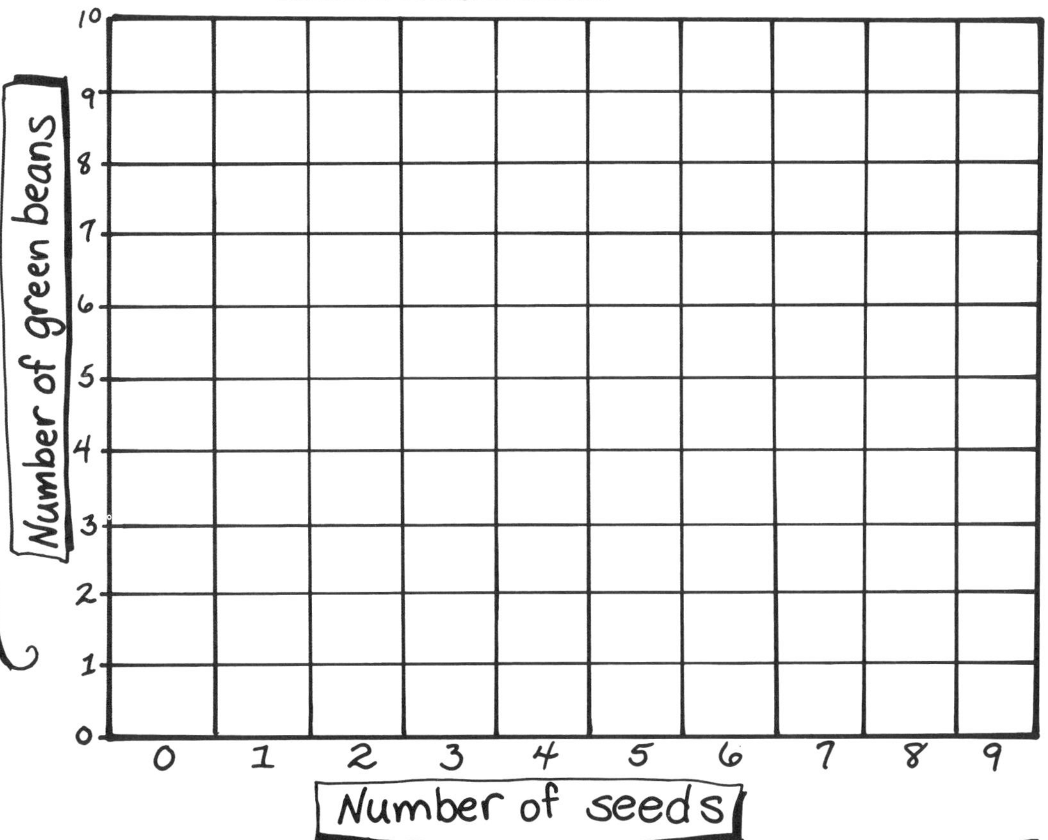

4. What is the most common number of seeds you found? ☐

Level 3   Student Name: ______________________________ Date: ______________

HYPOTHESIZING

Materials

Level 3 Hypothesizing Assessment Sheet
20 gram weight cubes
ruler, pencil, tape

Procedure

Explain that a hypothesis is a testable statement. It states what you are going to do and what you expect to happen. Demonstrate how to tape a small cup to each end of the ruler and place the pencil under the middle of the ruler at the 15 centimeter mark. Then have students place 3 gram weight cubes in one of the cups (the load). Ask students to put cubes into the cup on the opposite end of the ruler until the lever (ruler) raises the load. Have students record how many cubes it takes to lift the load (the effort). Then ask students to complete the hypotheses: If the fulcrum (pencil) is moved closer to the load, then the effort to move it will **increase** or **decrease**; and If the fulcrum (pencil) s moved farther away from the load, then the effort to move it will **increase** or **decrease**, Have students test their hypotheses by moving the fulcrum first to 8 centimeters from the load and then to 15 centimeters, and finally to 23 centimeters while placing gram weight cubes in the other cup until the load of 3 gram weight cubes are lifted each time. Ask students to record the number of gram weights needed to raise the loads in the table and explain if their results supported their hypotheses.

Hypothesizing Rubric

4 points   exceptionally accurate/reasonable, complete, detailed
3 points   adequately accurate/reasonable, complete, detailed
2 points   somewhat accurate/reasonable, complete, detailed
1 point   not accurate/reasonable, complete, or detailed
0 points   not attempted

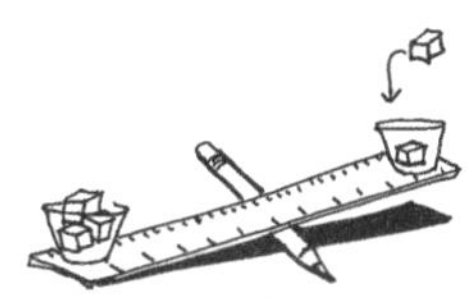

Scoring Criteria- Estimating	Points 0-4
Followed directions using safe procedures to complete the performance assessment.	
Successfully set up the lever with materials, ruler, pencil, cups.	
Placed 3 gram weights in one cup, added gram weights to the other cup until the load was lifted, and recorded the number of gram weights used to raise the load.	
Completed the first hypothesis by writing whether moving the fulcrum closer to the load will increase or decrease the effort needed to raise the load.	
Completed the second hypothesis by writing whether moving the fulcrum farther away from the load will increase or decrease the effort needed to raise the load.	
Recorded the number of gram weights to lift the load of 3 gram weights when the fulcrum is at 8 centimeters, 15 centimeters, and 23 centimeters from the load.	
Explained whether or not the results supported the hypotheses.	
Total	/24

Comments:

Hypothesizing

How does the position of the fulcrum affect the amount of effort needed to lift the load?

1. Make a lever with a ruler and a pencil.
   - Tape a small cup to each end of the ruler.
   - Put the pencil under the middle at the 15 centimeter mark.

2. Place 3 cubes in one of the cups. This is the load. Add cubes to the other side until the lever raises the load.

   ★How much effort did it take to lift the load? [ ] cubes

3. If the fulcrum (pencil) is moved closer to the load, then the effort to move it will ________________.
   (increase or decrease)

4. If the fulcrum (pencil) is moved farther away from the load, then the effort to move it will ________________.
   (increase or decrease)

5. Test your hypothesis. Record your results.

Fulcrum distance from load	Load	Effort needed to lift load
8 centimeters	3 cubes	cubes
15 centimeters	3 cubes	cubes
23 centimeters	3 cubes	cubes

6. Did your results support your hypothesis? Explain.

____

____

Level 3 Student Name: ______________________________ Date: ______________

DEFINING OPERATIONALLY

Materials

Level 3 Defining Operationally Assessment Sheet
Baking Soda,
6 Plastic cups
Liquids: Vinegar, water, orange juice, lemon juice, fruit juice, your choice

Procedure

Tell students that when they define operationally, they tell what they did and what happened. Show students the baking soda and plastic cups of liquid (vinegar, water, orange juice, lemon juice, fruit juice, and another liquid). Instruct students to put a pinch of baking soda into the vinegar and describe what happens. Tell students that baking soda is an indicator of acid and the reaction they observed (fizzing) indicated that vinegar is an acid. Have students put a pinch of baking soda in each of the liquids and describe the reaction. Ask them to write "yes" or "no" to indicate whether or not the liquid is an acid based on the reaction they observed with baking soda (fizzing or no fizzing). Then have students give an operational definition of acid based on what they did and what happened.

Defining Operationally

4 points — exceptionally accurate/reasonable, complete, detailed
3 points — adequately accurate/reasonable, complete, detailed
2 points — somewhat accurate/reasonable, complete, detailed
1 point — not accurate/reasonable, complete, or detailed
0 points — not attempted

Scoring Criteria- Defining Operationally	Points 0-4
Followed directions using safe procedures to complete the performance assessment.	
Added a pinch of baking soda to vinegar and drew and described what happened.	
Added a pinch of baking soda to water, described what happened, and wrote "yes" or "no" to indicate if water is an acid.	
Added a pinch of baking soda to orange juice, described what happened, and wrote "yes" or "no" to indicate if orange juice is an acid.	
Added a pinch of baking soda to lemon juice, described what happened, and wrote "yes" or "no" to indicate if lemon juice is an acid.	
Added a pinch of baking soda to fruit juice, described what happened, and wrote "yes" or "no" to indicate if fruit juice is an acid.	
Added a pinch of baking soda to another liquid, described what happened, and wrote "yes" or "no" to indicate if the other liquid is an acid.	
Gave an operational definition of "acid" based on what was done and what happened.	
Total	/ 32

Comments:

Defining Operationally

1. Place a pinch of baking soda into some vinegar.
Draw and describe what happens.

____

____

Baking soda is an indicator of acid. Vinegar has acid and is sour or bitter to the taste.

2. Place a pinch of baking soda in each of the liquids to test for acid.

Liquid	Describe reaction with Baking Soda	Acid? yes or no
water		
orange juice		
lemon juice		
fruit juice		

3. Give your definition of an acid based on what you did in this activity.

____

____

____

Level 3 Student Name ______________________________ Date ______________

CONTROLLING VARIABLES

Materials

Level 3 Controlling Variables Assessment Sheet
ruler with grove down the middle
marble, metric tape measure, 4 books

Procedure

Ask students to use the word variable in a sentence that tells something about science. After several students have given sentences, ask what is meant by, controlling variables. Ask students why controlling variables is important in conducting science investigations. Then tell students that they will have an opportunity to control variables in a science investigation. Demonstrate how to make a ramp by setting a ruler on top of a book, how to roll a marble from the top of a ramp, and how to use a tape measure to measure the distance in centimeters from the end of the ruler to the point where the marble stops rolling. Instruct students to repeat this procedure adding a second book, third book, and fourth book. Then explain that they will think about the variables in the investigation and determine the variable that was changed, the variable that responded to the change, and the variables that were kept the same.

Controlling Variables Rubric

4 points	exceptionally accurate/reasonable, complete, detailed
3 points	adequately accurate/reasonable, complete, detailed
2 points	somewhat accurate/reasonable, complete, detailed
1 point	not accurate/reasonable, complete, or detailed
0 points	not attempted

Scoring Criteria- Controlling Variables	Points 0-4
Followed directions using safe procedures to complete the performance assessment.	
Rolled a marble down a ramp one book high, measured the distance the marble traveled from the ramp, and recorded the measurement in centimeters.	
Rolled a marble down a ramp two books high, measured the distance the marble traveled from the ramp, and recorded the measurement in centimeters.	
Rolled a marble down a ramp three books high, measured the distance the marble traveled from the ramp, and recorded the measurement in centimeters.	
Rolled a marble down a ramp four books high, measured the distance the marble traveled from the ramp, and recorded the measurement in centimeters.	
Listed the variable that was changed in the investigation.	
Listed the variable that responded to the change in the investigation.	
Listed at least three variables that were kept the same in the investigation.	
Total	/32

Comments:

Controlling Variables

1. Make a ramp by setting the ruler on top of one book.
2. Roll the marble from the top of the ramp and measure the distance the marble rolls from the end of the ruler. Record the distance.
3. Add another book to raise the ramp. Roll the marble again. Measure and record the distance it rolls.
4. Repeat this procedure until the ramp is <u>four</u> books high.

Height of Ramp	Distance the marble travels from end of ramp.
1 book	centimeters
2 books	centimeters
3 books	centimeters
4 books	centimeters

5. Which variable did you change? ____________________
6. Which variable responded to the changes? (What did you measure?) ____________________
7. Which variables did you keep constant? (What did you keep the same?) ____________________

____

Level 3 Student Name: ______________________________ Date: ______________

INVESTIGATING

Materials

Level 3 Investigating Assessment Sheet
Water, Vinegar, 6 Plastic Cups,
Eye dropper, waxed paper
Salt labeled Powder A
Sugar labeled Powder B
Flour labeled Powder C
Baking soda labeled Powder D

Procedure

Label the 6 plastic cups: Powder A (salt), Powder B (sugar), Powder C (flour), Powder D (baking soda), Water, and Vinegar. Put the corresponding substances into each of the labeled cups. Explain that there is a problem in the kitchen. The chef has mixed up the ingredients for a cake and needs to find out which white powder is baking soda. Ask students to design and conduct an investigation to find out which white powder is baking soda. Instruct them to construct a chart to show the results of their investigation. Then have them draw a conclusion about the results of their investigation and describe how they decided which powder was baking soda.

Investigating Rubric

4 points	exceptionally accurate/reasonable, complete, detailed
3 points	adequately accurate/reasonable, complete, detailed
2 points	somewhat accurate/reasonable, complete, detailed
1 point	not accurate/reasonable, complete, or detailed
0 points	not attempted

Scoring Criteria- Investigating	Points 0-4
Followed directions using safe procedures to complete the performance assessment.	
Described how to conduct an investigation to find out which white powder is baking soda.	
Described how Power A felt, how it reacted to water, and how it reacted to vinegar.	
Described how Power B felt, how it reacted to water, and how it reacted to vinegar.	
Described how Power C felt, how it reacted to water, and how it reacted to vinegar.	
Described how Power D felt, how it reacted to water, and how it reacted to vinegar.	
Drew a conclusion based on the results of the data collected.	
Described the evidence used to decide which power was baking soda.	
Total	/ 32

Comments:

Investigating

1. There is a problem in the kitchen. The chef has mixed up the ingredients for the cake. Which of these white powders is the baking soda?

2. Design and conduct an investigation to find the baking soda. Tell what you will do.

3. Construct a chart to show your results.

Powder	Feel	Reaction to Water	Reaction to Vinegar
A			
B			
C			
D			

4. Conclusion: Which of these powders is baking soda?

5. How did you decide which powder is baking soda?

SCIENCE INVESTIGATIONS

TEMPLATE for TEACHING and ASSESSING

National Science Education Content Standards Science as Inquiry

- Abilities necessary to do scientific inquiry
- Understandings about scientific inquiry

Use this template to guide students to plan and conduct their own science inquiry investigations generated from questions they ask and answer. Each time you have students conduct a science activity designed to help them understand a science concept, ask students what questions they have.

These questions can be written on the chalkboard/whiteboard or posted on a bulletin board. Then have students select an investigable question by guiding students to eliminate questions that may present safety issues or be impractical to do in the classroom because of equipment, space, or time constraints. Read each question and then ask students if they can answer the question safely in a limited amount of time using the materials available in the classroom. Then eliminate the "Why?" and "How?" questions. Explain to students that although these questions are very important and often drive scientific research, we may not be able to answer the "Why?" and "How?" questions by conducting a classroom investigation.

Explain that investigations that can be conducted in the classroom include: *systematic observations (descriptive investigations)* in which questions are answered by making careful observations and/or measurements or *experiments* in which comparisons are made. Tell students that comparison questions can be answered by conducting an **experiment** in which there is a manipulated/independent variable (what is changed), a responding/independent variable (what is observed or measured), and controlled variables (what is kept the same). Explain that **systematic observations (descriptive investigations)** require the same careful planning by setting the conditions for making observations and/or measurements.

Clarify that if students are answering a "what-would-happen-if" or "measuring" question, they will conduct an investigation that is a **systematic observation (descriptive investigation)** in which they predict the answer to their question. A prediction is an educated guess based on observations, inferences, and prior knowledge that are used to develop an idea of an expected result. Explain that if they are making a "comparison" they construct a hypothesis, a statement about the relationship between the independent (manipulated) and dependent (responding) variables that can be tested experimentally.

Tell students that the next step is to plan a procedure for testing their prediction or hypothesis. Emphasize that it is important to think about all the variables (things that can be changed) in a **systematic observation** (**descriptive investigation**) and an **experiment**. Explain that if the question can be answered by doing a **systematic observation** (**descriptive investigation**) that students should write a step-by-step procedure listing exact amounts and measurements they will use to conduct their investigations and collect data. If the question can be answered by doing an **experiment,** they should list the manipulated (what will be changed), responding (what will be observed or measured), and controlled variables (what will be kept the same).

Data collected can be presented in a chart, graph, drawing, description, photograph, etc. Data can then be analyzed to draw a conclusion to point out the trends patterns, and relationships indicated by the data collected. Tell students the next step is to give an explanation for their results in the conclusion.

Constructing explanations requires the use of a logical thought process to explain the relationships involving interactions, dependencies, and cause-and-effect between and among objects and events.

Then have students think about a question they had while they were conducting an investigation and write it under "Next Question." It has been said that good inquiry occurs when two questions grow where there was only one before!

Students can use the template to plan, conduct, and record the results of their own inquiry investigations. If students have the opportunity to ask and answer their own questions after each science investigation, they will be engaged in full scientific inquiry. It will not require more preparation of materials if you generate questions following each science activity and eliminate those that can't be answered with the materials that are at hand.

Have students fold the template in half (the long way) and draw a line on the resulting crease down the middle of the template. Then have students cut on the thin solid lines and fold on the thick lines. This will form tabs under which students can write. Instruct students to draw lines horizontally between the cuts (they can fold the template and draw along the creases). Explain that students will write in the spaces defined by these lines under each tab.

Student Name ______________________ Date ____________

Inquiry Rubric

4 points — exceptionally accurate/reasonable, complete, detailed
3 points — relatively accurate/reasonable, complete, detailed
2 points — slightly accurate/reasonable, complete, detailed
1 point — not accurate/reasonable, complete, or detailed
0 points — not attempted

Scoring Criteria- Inquiry Rubric	Points 0-4
Question can be answered with a simple investigation	
Guess suggests a logical answer to the question.	
Listed or made drawings of each step required to carry out the investigation.	
Described observations and/or measurements using drawings and/or words.	
Described results in words or drawings and compared results to guess.	
Answer for the question was based on actual results	
Wrote a follow up question that could be investigated based on observations during investigation	
Total	/28

Comments

The template includes the scoring rubric on the back so students know what the expectations are for the inquiry investigation and teachers can easily score the plan and the inquiry project.

If you keep a bulletin board in you classroom with questions generated after each science activity, students can select one of the questions from the bulletin board for their science fair projects. Science fair projects will then extend from the student-generated questions as they are learning science concepts. With practice after each science investigation using this template, students will have less difficulty coming up with a question and a plan to answer that question for their science fair project.

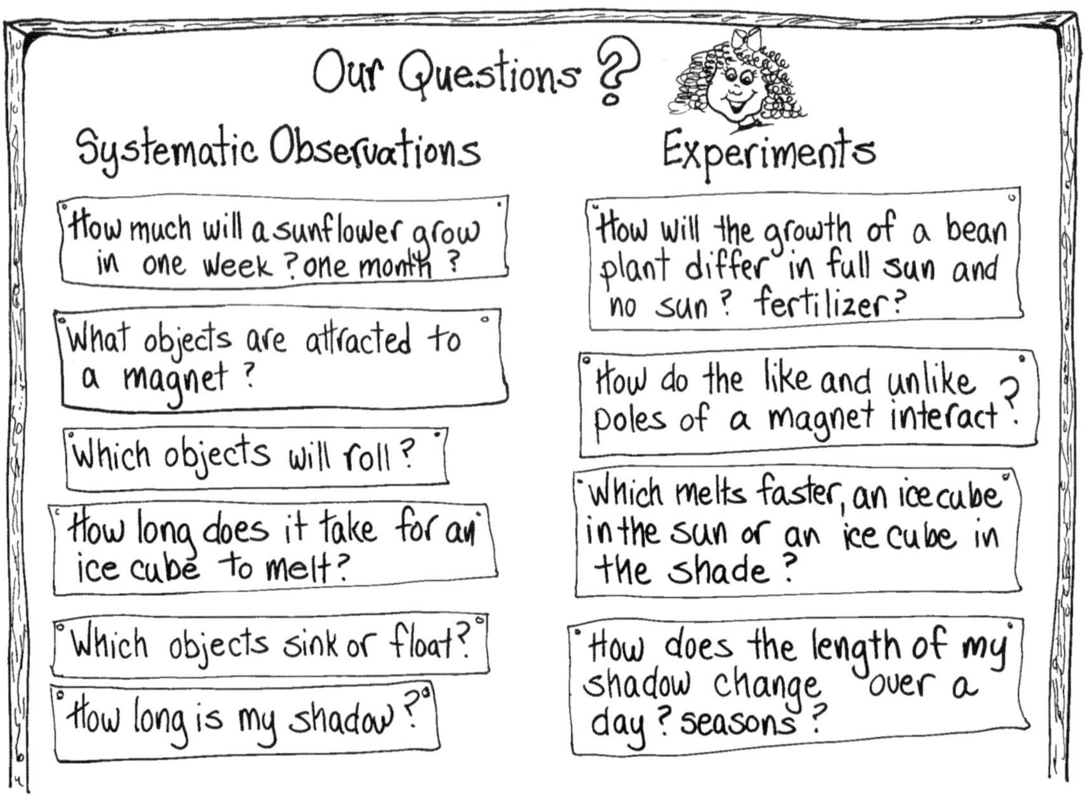

My Guess

My Materials

My Results

My Next Question

Student Name:______________________ Date:__________

Inquiry Rubric

4 points	exceptionally accurate/reasonable, complete, detailed
3 points	accurate/reasonable, complete, detailed
2 points	slightly accurate/reasonable, complete, detailed
1 point	not accurate/reasonable, complete, or detailed
0 points	not attempted

Scoring Criteria- Inquiry	Points 0-4
Question can be answered with a simple investigation	
Guess suggests a logical answer to the question.	
Listed or made drawings of each step required to carry out the investigation.	
Described observations and/or measurements using drawings and/or words.	
Described results in words or drawings and compared results to guess.	
Answer for the question was based on actual results	
Wrote a follow up question that could be investigated based on observations during investigation	
Total	/ 28

Comments:

My Question

My Plan

My Observations

My Answer to the Question

WHAT DO SCIENTISTS DO?

Scientists conduct scientific investigations by asking and answering a question.

Scientists use different kinds of investigations depending on the question they are trying to answer.

- *What would happen if* questions and *measuring* questions are answered by conducting **systematic observations**.
- *Comparison* questions are answered by using a fair test to conduct an **experiment.**

Scientists use instruments like magnifiers, rulers, and thermometers to help them obtain more information than they obtain by using only their senses.

Scientists develop explanations using observations (evidence) and what they already know about the world (scientific knowledge).

Scientists make the results of their investigation public; they describe the investigation in ways that enable others to repeat the Investigation.

Scientists review and ask questions about the results of other scientists' work.

I ♥ Science
Participant
I ♥ Science
Excellent
I ♥ Science
Superior

I ♥ Science !
SAHeir 2006